DONALD LLOYD McKINLEY
1460 S. Sheridan Way
Mississauga, Ontario L5H 1Z7

Polyhedra Primer

$12.00

DONALD LLOYD McKINLEY
Designer - Craftsman
346 Erin Street
Oakville, Ontario   L6H 4P9

Polyhedra Primer

Peter Pearce and Susan Pearce

VNR VAN NOSTRAND REINHOLD COMPANY
NEW YORK   CINCINNATI   TORONTO   LONDON   MELBOURNE

Copyright© 1978 by Peter Pearce and Susan Pearce

Library of Congress Catalog Card Number: 78-1626

ISBN 0-442-26496-8

All rights reserved. No part of this work covered by the copyright hereon may be reproduced or used in any form or by any means—graphic, electronic, or mechanical, including photocopying, recording, taping, or information storage and retrieval systems—without written permission of the publisher and the authors.

Printed in the United States of America

Published in 1978 by Van Nostrand Reinhold Company
A division of Litton Educational Publishing, Inc.
135 West 50th Street, New York, NY 10020, U.S.A.

Van Nostrand Reinhold Limited
1410 Birchmount Road
Scarborough, Ontario M1P 2E7, Canada

Van Nostrand Reinhold Australia Pty. Ltd.
17 Queen Street
Mitcham, Victoria 3132, Australia

Van Nostrand Reinhold Company Limited
Molly Millars Lane
Wokingham, Berkshire, England

16　15　14　13　12　11　10　9　8　7　6　5　4　3　2　1

LIBRARY OF CONGRESS CATALOGING IN PUBLICATION DATA

Pearce, Peter.
　Polyhedra primer.

　Bibliography: p.
　Includes index.
　1. Polyhedra.　I. Pearce, Susan, joint author
II. Title
QA491.P4　　　　516'.15　　　　78-1626
ISBN 0-442-26496-8

Illustrations and book design by Susan Pearce and Peter Pearce.

Contents

Preface vii
1. **Polygons** 1
2. **Tessellations** 21
3. **Polyhedra** 45
4. **Dual Polyhedra** 73
5. **Space Filling** 85
6. **Open Packings** 107
7. **Constructions** 117

Bibliography 131
Index 133

Preface

The idea that a book such as *Polyhedra Primer* would be a useful addition to the literature on the geometry of polyhedra arose after ten years of study and original work on the geometry of three-dimensional space and its application to building systems. These studies culminated in a book by Peter Pearce entitled *Structure in Nature Is a Strategy for Design* (MIT Press, 1978)—a large, profusely illustrated book that will attract but a small audience because of its price and the detailed scope of its subject matter.

During this period of study countless geometric models were built in many different media; numerous discussions were held with people knowledgeable in the field; and an extensive and thorough search of the literature on polyhedra was undertaken. It became clear that accurate, useable sources, other than some excellent, high-level, mathematically oriented material, was virtually nonexistent. After completing *Structure in Nature*, it seemed appropriate to produce a general interest, low-cost book that would be more accessible while maintaining high standards of rigor and thoroughness.

The three-dimensional spatial relationships of polyhedra are basically nonabstract and are particularly amenable to visual representation and communication. Such geometry, unlike other technical subjects, can be illustrated without compromising the rigor of the subject matter. *Polyhedra Primer* is thus conceived as a visual attempt to facilitate the understanding of the principles embodied in polyhedra.

The goal of *Polyhedra Primer* is to teach the geometry of polyhedra, and as such is a pedagogical presentation, not a philosophical treatise. It can serve as reference book while at the same time providing the reader with new ideas about the geometric organization of three-dimensional space. It is designed to enable the user to easily locate specific ideas and concepts. It is an illustrated glossary organized, not alphabetically, but in a hierarchical sequence from the simplest idea to the more complex. Because of this format, the general reader will be able to develop a basic understanding of the fundamental concepts without having to be a mathematician. For this reason, we hope that the book will fill a niche that is not presently covered by the literature in the field.

The book contains chapters on polygons, tessellations, finite polyhedra, space filling, and open packings. A chapter on the construction of geometric models is also included, because we believe that the physical manipulation of actual models is a useful learning experience.

Written to be understood at the high school and college level, *Polyhedra Primer* will appeal to both academic and general audiences. We think it will be especially meaningful for people in those professions—architecture, planning, engineering, industrial design, and art—where a knowledge of geometry provides a rich resource of form and spatial options and is useful as the basis for the formulation of new solutions to design problems. It is hoped that the book will find further audience among teachers, students, and practitioners of mathematics, crystallography, general morphology, and other areas of scientific endeavor that require a knowledge of spatial geometry.

We hope that beyond such pedagogical purposes, the reader will enjoy our little book and perhaps will be struck, as we have been, by the extraordinary spatial diversity that emanates from the sublime order and elegant simplicity that is exemplified by the subject of polyhedra.

1 Polygons

Points, Lines, and Angles

**Points
Line**

Two points can be connected by a straight line on a flat surface or plane.

Parallel

Lines are parallel if they lie in a common plane and do not intersect no matter how far they are extended.

Angle

An angle is the figure formed by two lines meeting at the same point.

**Degrees
Minutes**

Angles are measured in degrees and minutes. There are 60 minutes in one degree. The symbol for degrees is (°). The symbol for minutes is (').

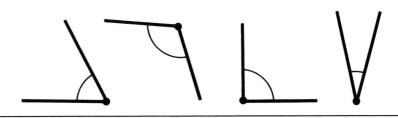

There are 360° 0' in a full circle.   There are 270° 0' in ¾ circle.

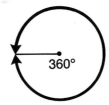

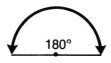

There are 180° 0' in ½ circle.   There are 90° 0' in ¼ circle.

Right Angle

A right angle has 90°.

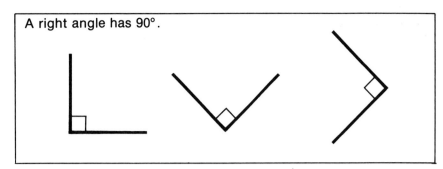

Acute Angle

An acute angle has less than 90°.

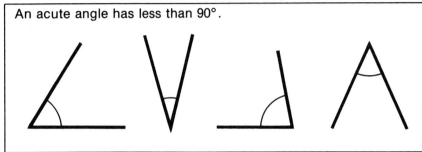

Obtuse Angle

An obtuse angle has more than 90° but less than 180°.

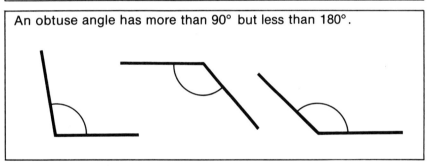

REFLEX ∠ = OVER 180°

Polygons

Three or more points can be connected by a line.

Three or more points can be connected by a line to form a closed loop. The closed loop is a polygon. A polygon is a portion of a plane bounded by three or more lines or segments.

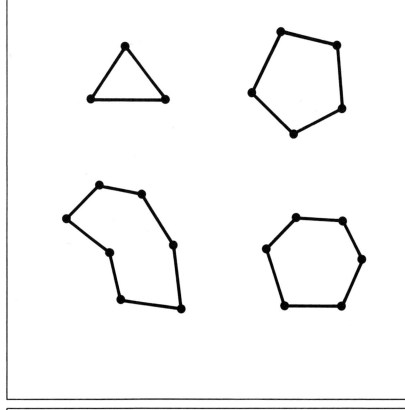

Vertex
Sides

A vertex is a corner of a polygon. (Plural: vertices). The sides of a polygon are the segments connecting the vertices.

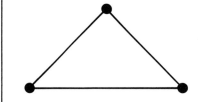 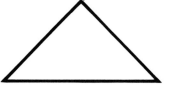

Interior

The interior of the polygon is the plane area bounded by the sides.

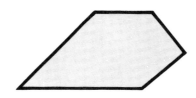 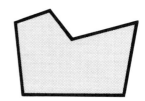

Interior or Face Angles

The interior or face angles are the angles formed by adjacent sides of a polygon and lying within the interior of the polygon.

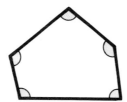 

If n is the number of sides of a polygon, the sum of the interior angles of a polygon is (n-2) × 180°.

 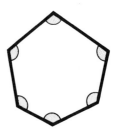

(5-2) × 180° = 540°   (6-2) × 180° = 720°

Convex Polygon

A polygon is convex if each interior angle is less than 180°.

Concave Polygon

A polygon is concave if one of its interior angles is more than 180°.

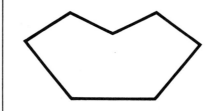 

Congruent

Polygons are congruent if they are the same shape and size. Congruent polygons will match exactly when placed one on top of the other.

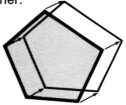 

Enantiomorph

A polygon is an enantiomorph if it exists in a left- and right-hand version. Enantiomorphs have all the properties of congruence except for handedness.

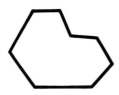 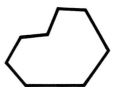

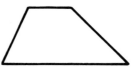 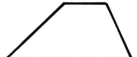

Naming Polygons　　　**n-gon**

Polygons are usually named by the number of sides they have. An n-gon is a polygon with an unspecified number of sides.

A *tri*angle has three sides.

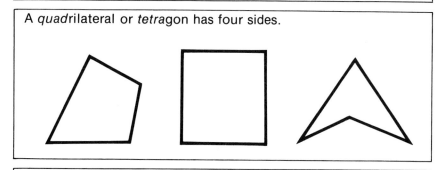

A *quad*rilateral or *tetra*gon has four sides.

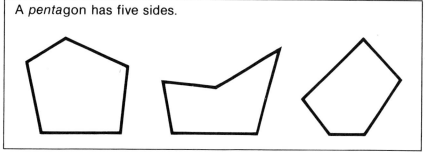

A *penta*gon has five sides.

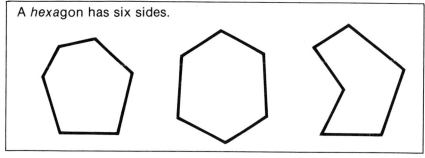

A *hexa*gon has six sides.

A *septa*gon has seven sides.

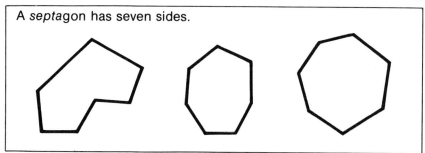

An *octa*gon has eight sides.

A *nona*gon has nine sides.

A *deca*gon has ten sides.

An *ennea*gon has eleven sides.

A *dodeca*gon has twelve sides.

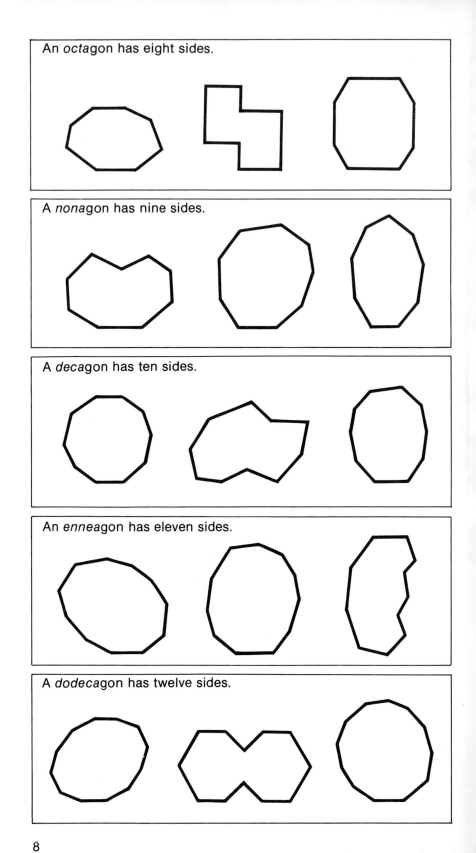

Properties of Polygons

Regular Polygons

A regular polygon has equal interior angles and equal sides.

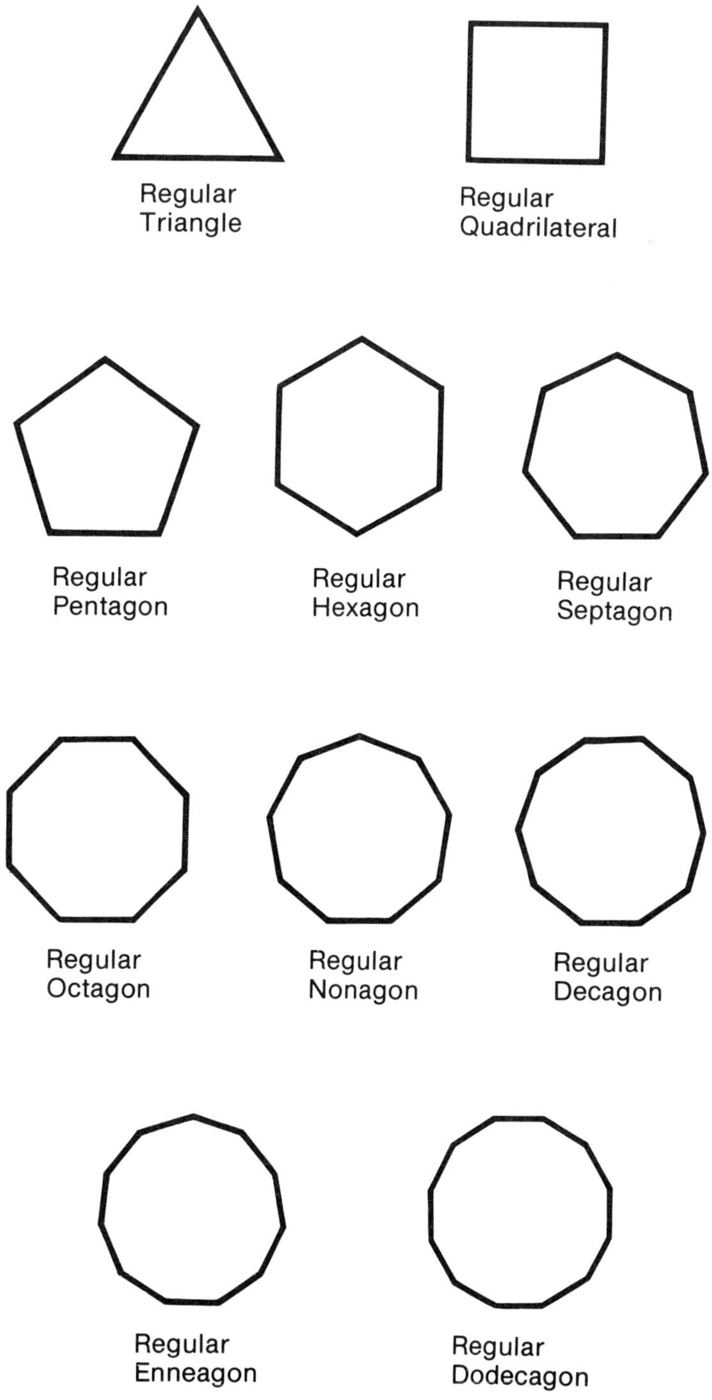

Regular Triangle

Regular Quadrilateral

Regular Pentagon

Regular Hexagon

Regular Septagon

Regular Octagon

Regular Nonagon

Regular Decagon

Regular Enneagon

Regular Dodecagon

Nonregular Polygons

A polygon can have equal interior angles and unequal sides.

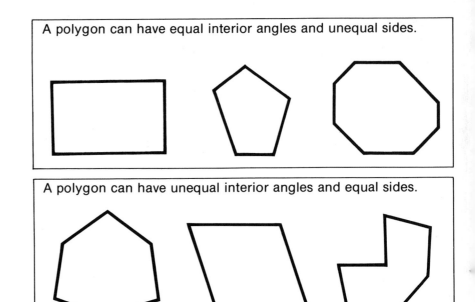

A polygon can have unequal interior angles and equal sides.

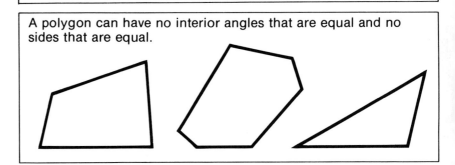

A polygon can have no interior angles that are equal and no sides that are equal.

Types of Triangles

The sum of the face angles of any triangle is always equal to 180°. (3-2) × 180° = 180°.

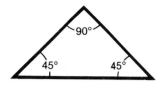 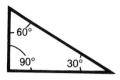

Equilateral

Equilateral triangles have equal length sides and equal interior angles. They are regular polygons.

Isosceles

Isosceles triangles have two sides of equal length and a third side—the base—of different length. The two interior angles common to the base are equal.

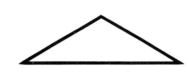 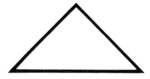

Scalene

Scalene triangles have three sides of different length and three different interior angles.

Right

Right triangles have one right (90°) angle. They can be isosceles or scalene. A right isosceles triangle has one right angle and two 45° angles.

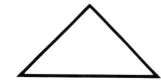 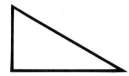

Types of Quadrilaterals

Trapezium

A trapezium has no parallel sides.

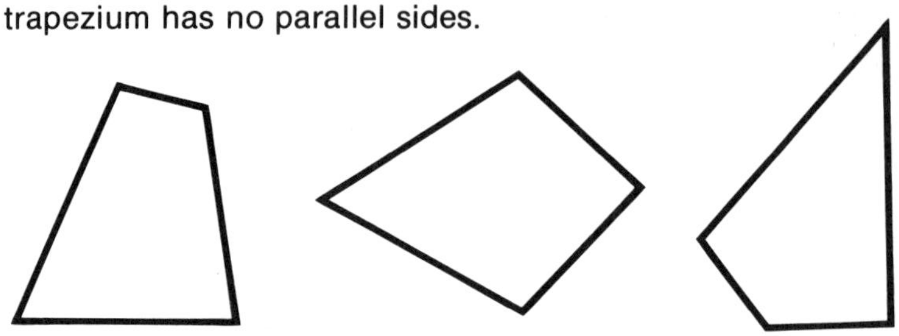

Trapezoid

A trapezoid has two parallel sides.

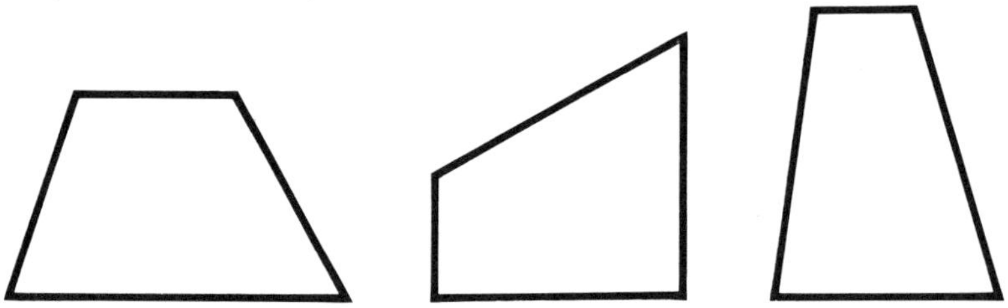

Parallelogram

A parallelogram has two pairs of parallel sides.

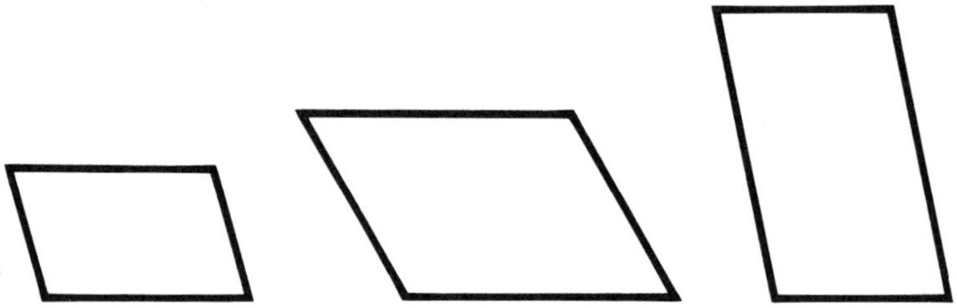

Rectangle

A rectangle is a parallelogram with four right angles.

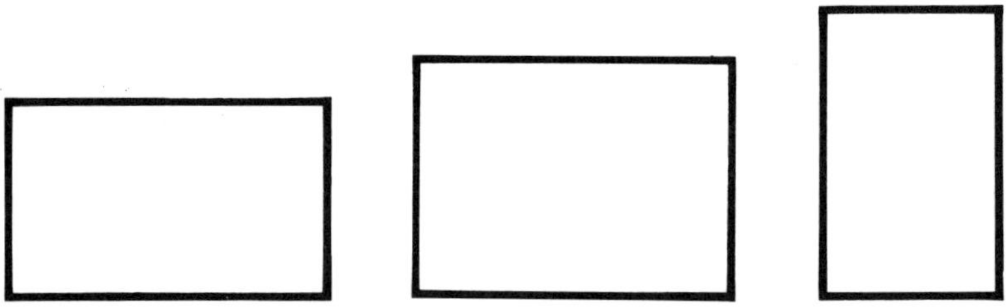

Rhombus

A rhombus is a parallelogram with equal sides.

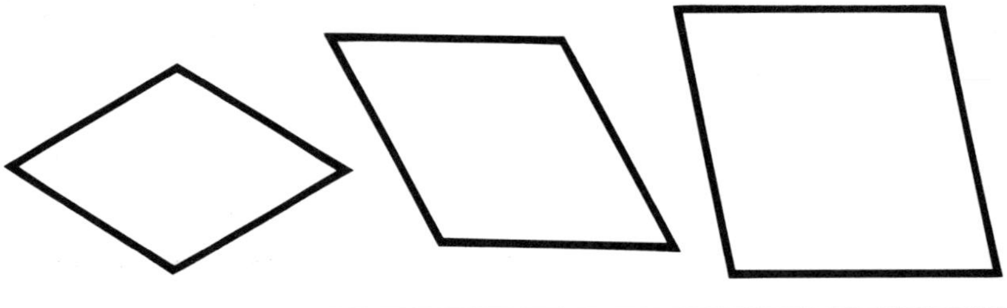

Square

A square is a parallelogram with equal sides and four right angles. It is a regular polygon.

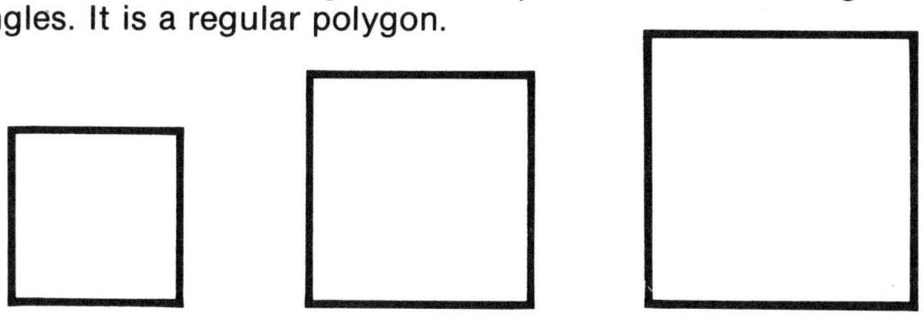

Symmetry and Polygons

Mirror Symmetry

A polygon has mirror symmetry, or one mirror plane, if it is the same on either side of a line that divides it in half.

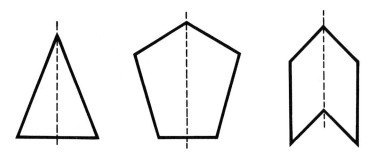

If a polygon with mirror symmetry is folded in half, the two folded parts will match exactly.

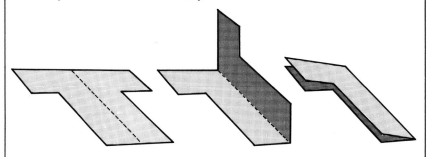

Or, if one half of a polygon with mirror symmetry is held up to a mirror, its reflection will look the same.

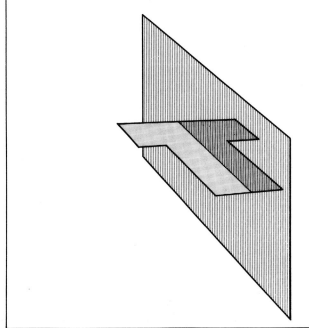

Rotational Symmetry

A figure has rotational symmetry if it repeats itself in one 360° revolution about an axis.

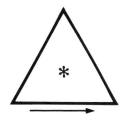

2-fold

A polygon has 2-fold rotational symmetry if it repeats itself twice in one 360° revolution about an axis.

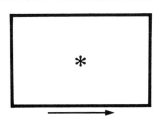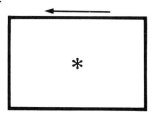

3-fold

A polygon has 3-fold rotational symmetry if it repeats itself three times in one 360° revolution about an axis.

4-fold

A polygon has 4-fold rotational symmetry if it repeats itself four times in one 360° revolution about an axis.

5-fold

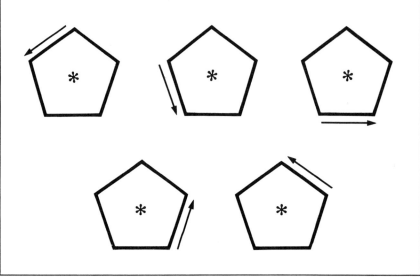

A polygon has 5-fold rotational symmetry if it repeats itself five times in one 360° revolution about an axis.

6-fold
n-fold

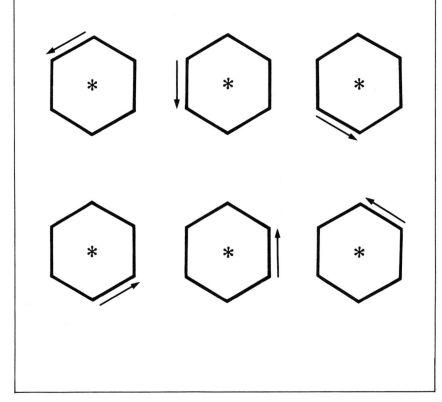

A polygon has 6-fold rotational symmetry if it repeats itself six times in one 360° revolution about an axis.
A polygon has n-fold rotational symmetry if it repeats itself n times in one 360° revolution about an axis.

Combinations of Symmetries

A polygon can have mirror symmetry and no rotational symmetry.

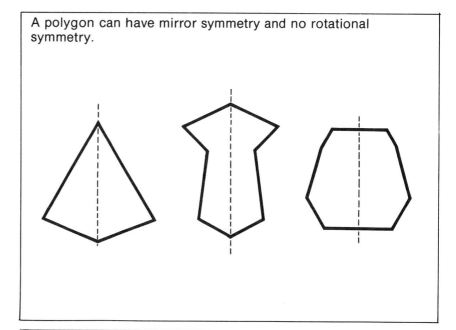

A polygon can have n-fold rotational symmetry and no mirror symmetry. In this case, the polygon will always exist as an enantiomorph.

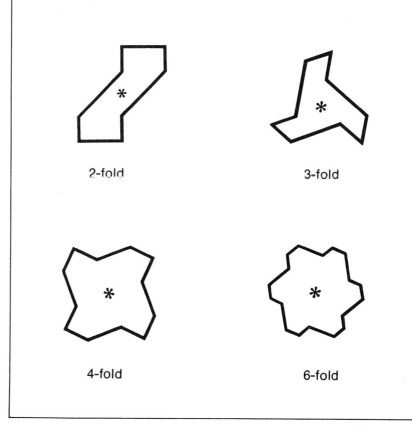

A polygon can have mirror symmetry and n-fold rotational symmetry. In this case, the polygon will always have the same number of mirror planes as n-fold symmetry.

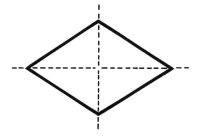

2-fold
2 mirror planes

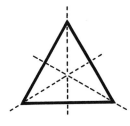

3-fold
3 mirror planes

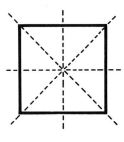

4-fold
4 mirror planes

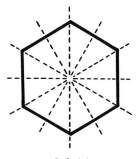

6-fold
6 mirror planes

Subdividing Polygons

Regular polygons can be subdivided into smaller congruent or enantiomorphic polygons.

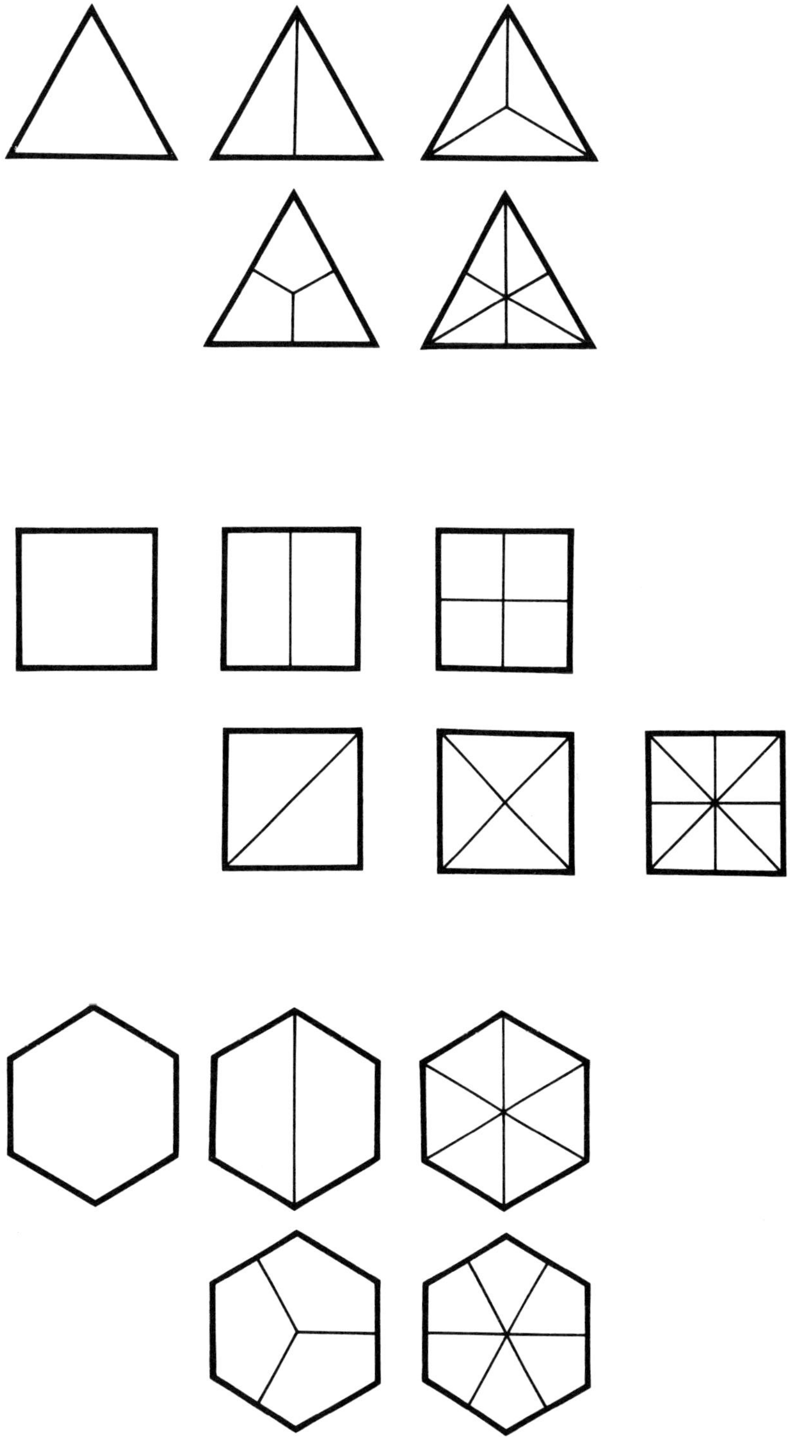

If the edges of a regular polygon are bisected, or cut in half, and the points of bisection joined between adjacent edges, smaller and smaller polygons of the same shape will result.

Star Polygons

Star polygons can be formed from polygons of more than five sides either by connecting alternate vertices of the polygon, or by extending the sides of the polygon until they intersect.

 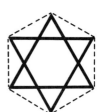

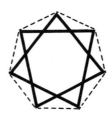 

Truncating Polygons

A truncated polygon is one whose corners have been cut off. Truncation forms a new polygon with more sides than the original polygon.

A regular polygon can be truncated to form a new regular polygon.

 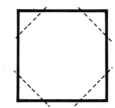 

2 Tessellations

Combining Polygons

In the plane, polygons can be joined together along matching edges.

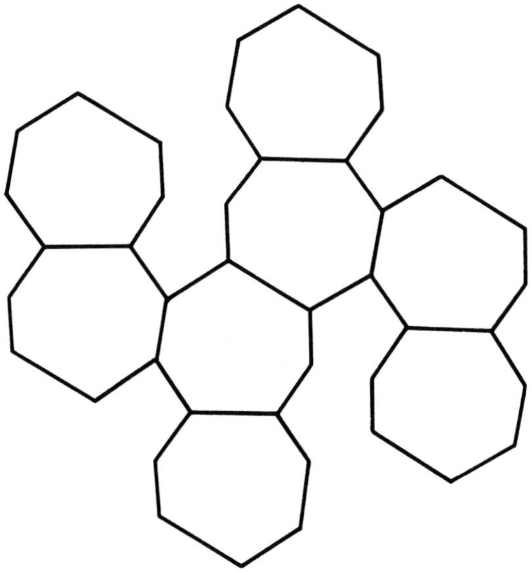

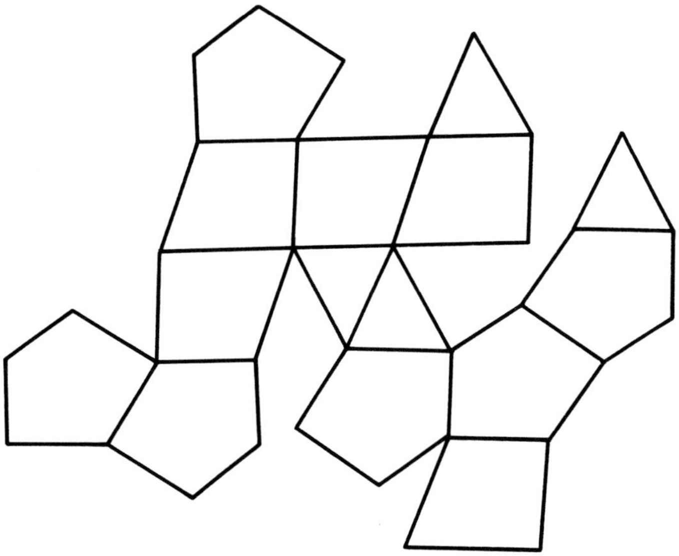

In the plane, the angle between joined polygons is 180°.

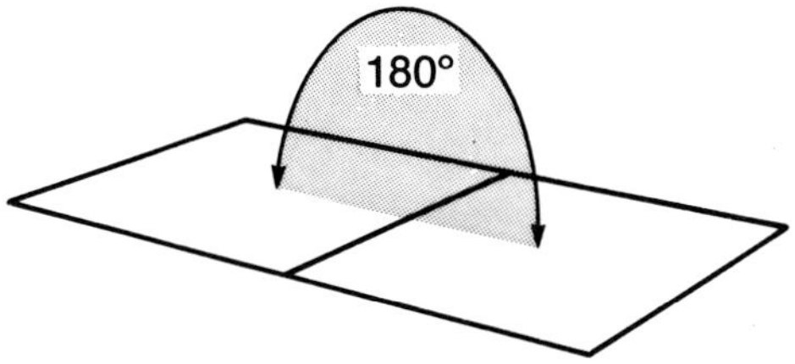

Tile Tessellation

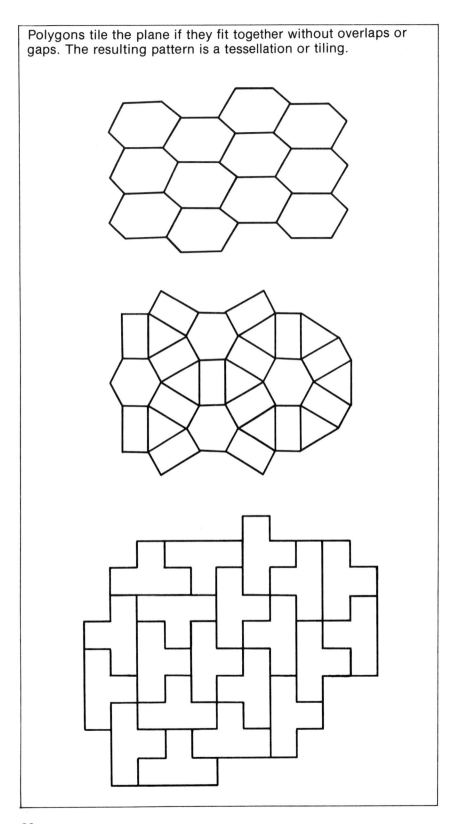

Vertex

In a tessellation, a vertex is formed when three or more sides of the polygons meet at a single point. Three sides meeting at a vertex is the minimum condition for tiling.

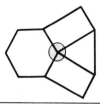 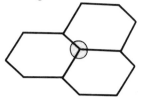

In a tessellation, the sum of the angles around each vertex is always equal to 360°.

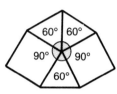 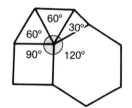

Uniform Tiling

In a uniform tiling, all vertices are congruent. That is, the arrangement of polygons around every vertex is the same.

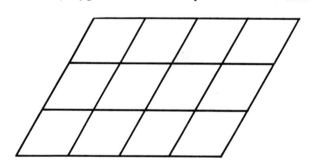

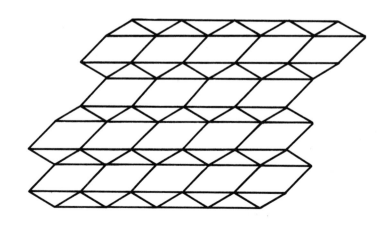

Periodic Tiling

In a periodic tiling, any region can be shifted to a new position where it again fits exactly. This region may be moved in any direction as long as it is not rotated.

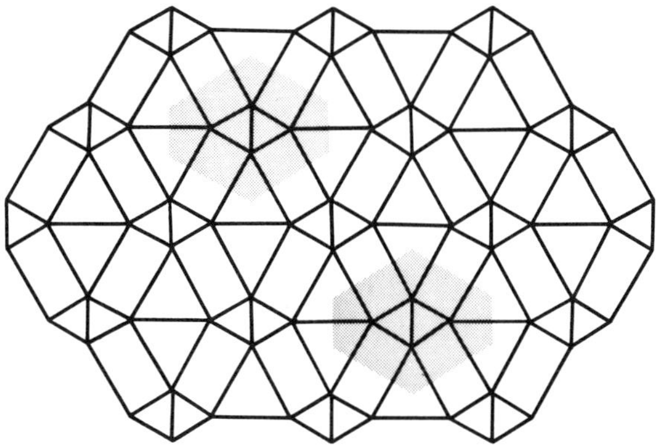

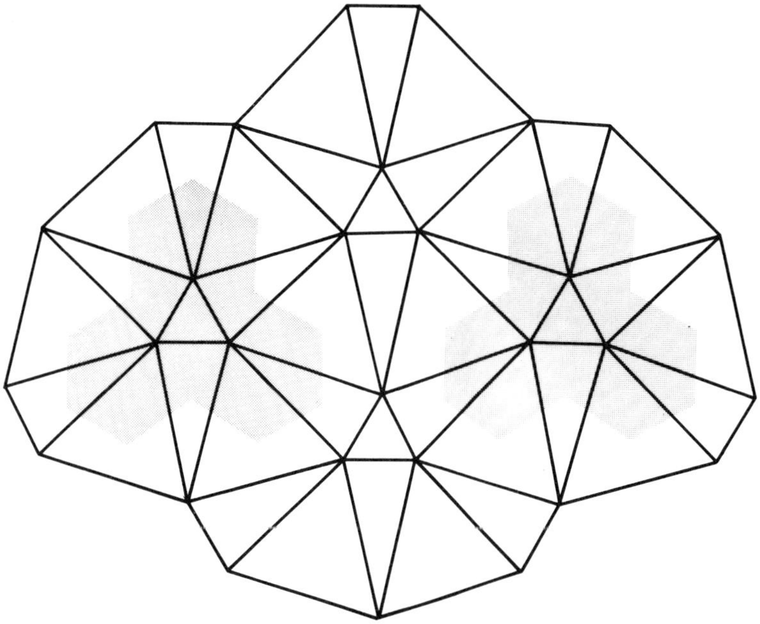

Tiling with Regular Polygons

Typical Vertices

Of the regular polygons, only triangles, squares, hexagons, octagons, and dodecagons can tile in various combinations around a common vertex. There are only 14 such combinations or typical vertices.

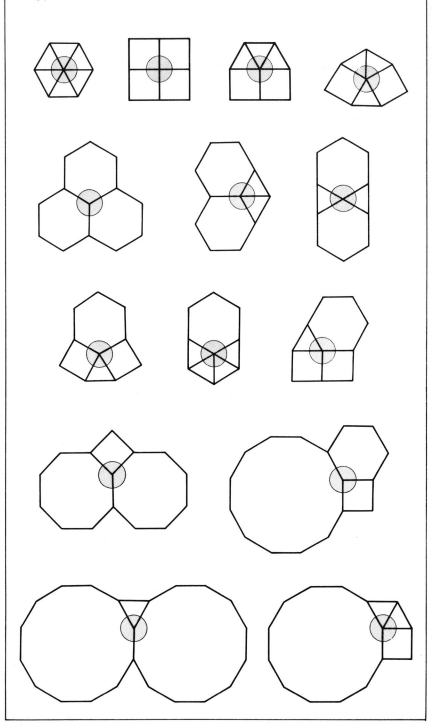

Regular Tessellations

A regular tessellation is periodic and uniform and consists of congruent regular polygons.

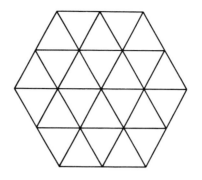 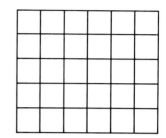

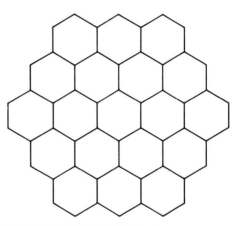

There are only three regular tessellations because there are only three regular polygons whose face angles divide evenly into 360°.

Equilateral triangles and squares can be repeated to form larger versions of themselves.

Semiregular Tessellations

A semiregular tessellation is periodic and uniform and consists of more than one kind of regular polygon. There are eight semiregular tessellations. One of them is an enantiomorph.

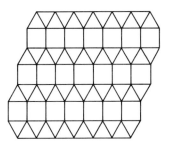
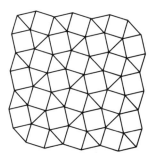
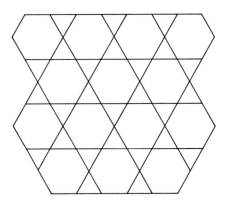
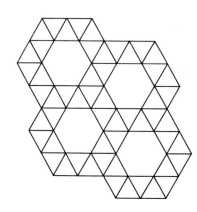
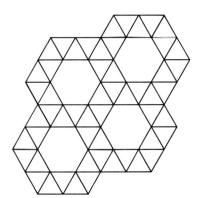

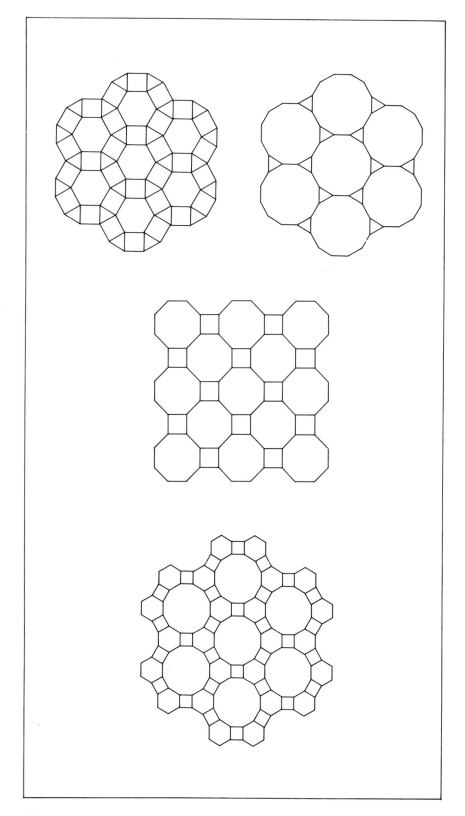

Nonuniform Periodic Tessellations with Regular Polygons

There is an infinite number of nonuniform periodic tessellations with regular polygons. However, because 360° is required around each vertex, we are still restricted to the 14 typical vertices for regular polygons.

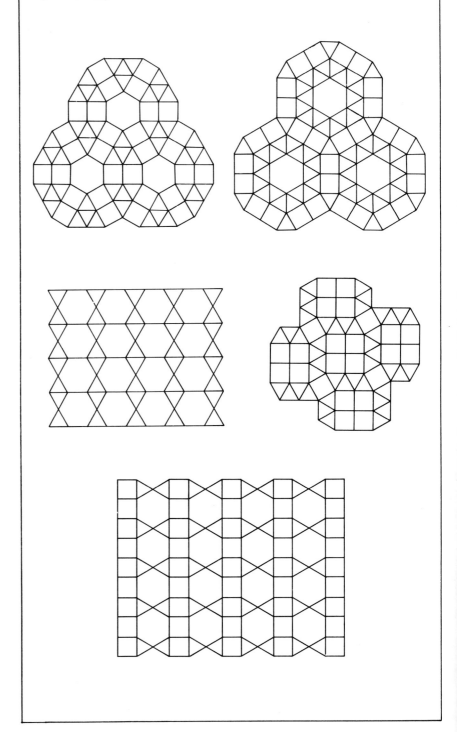

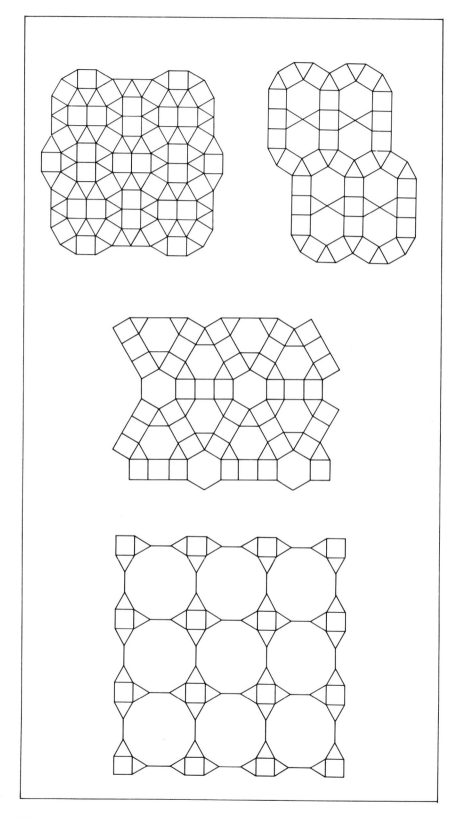

Nonuniform Nonperiodic Tessellations with Regular Polygons

There is an infinite number of nonuniform nonperiodic tessellations with regular polygons. Again, we are restricted to the 14 typical vertices.

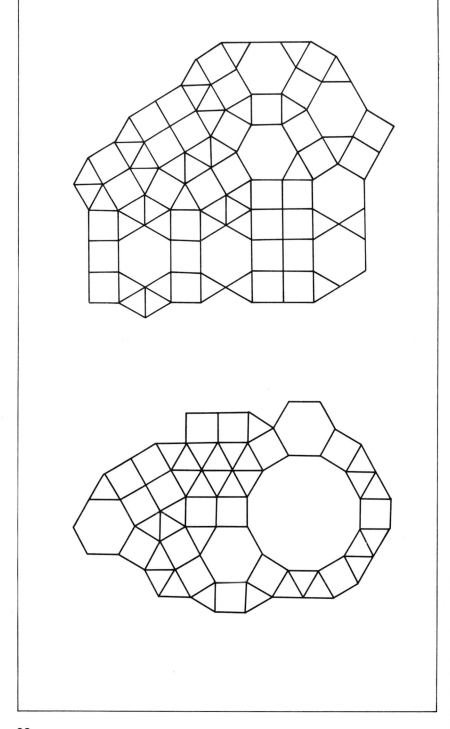

Tiling with Nonregular Polygons

Any triangle tiles the plane.

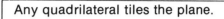

Any quadrilateral tiles the plane.

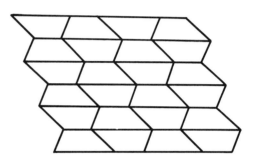

Any hexagon with three sets of equal parallel sides tiles the plane.

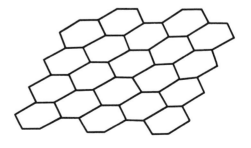

Uniform Periodic Tessellations with Nonregular Polygons

There is an infinite number of uniform periodic tessellations with nonregular polygons.

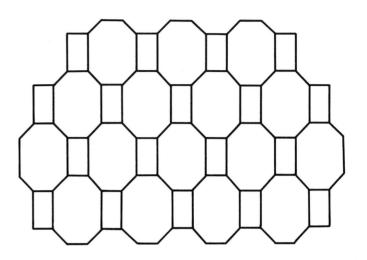

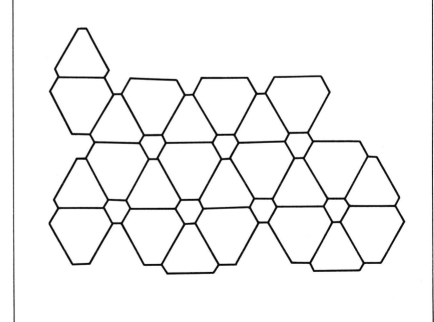

Nonuniform Periodic Tessellations with Nonregular Polygons

There is an infinite number of nonuniform periodic tessellations with nonregular polygons.

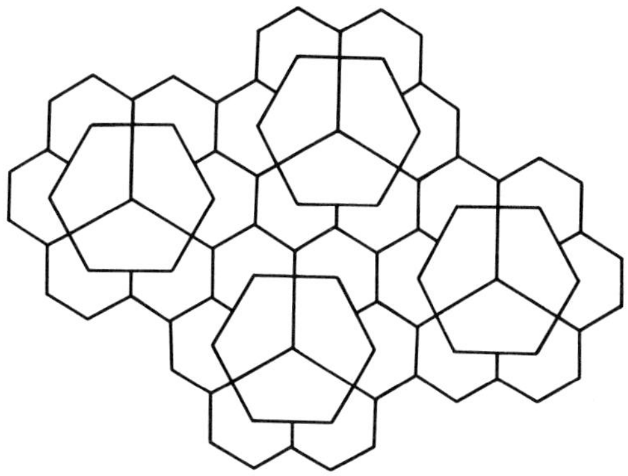

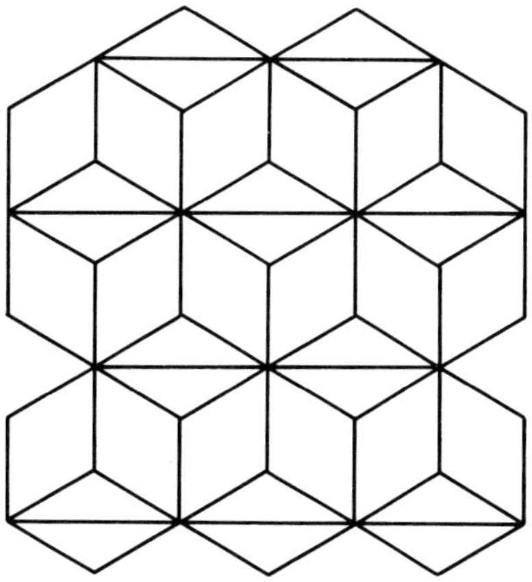

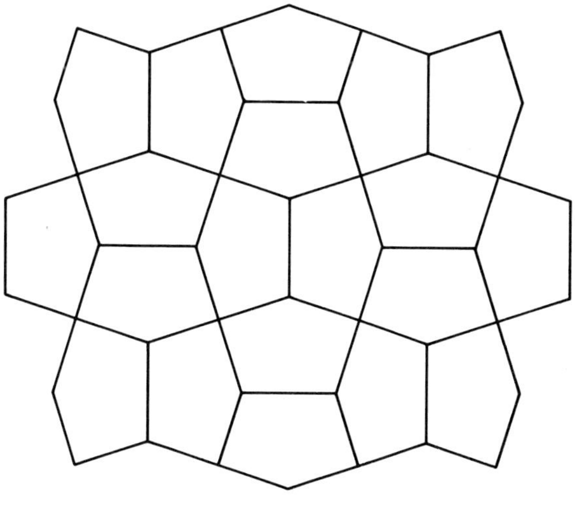

Nonuniform Nonperiodic Tessellations with Nonregular Polygons

There is an infinite number of nonuniform nonperiodic tessellations with nonregular polygons.

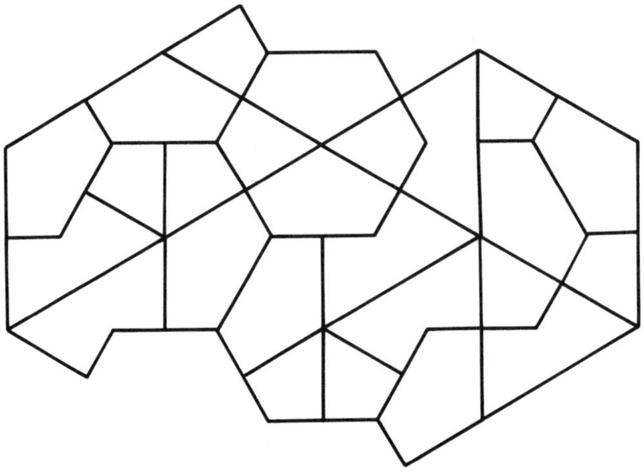

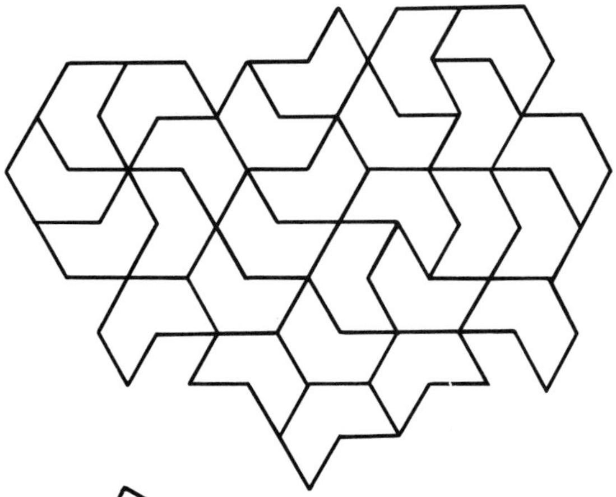

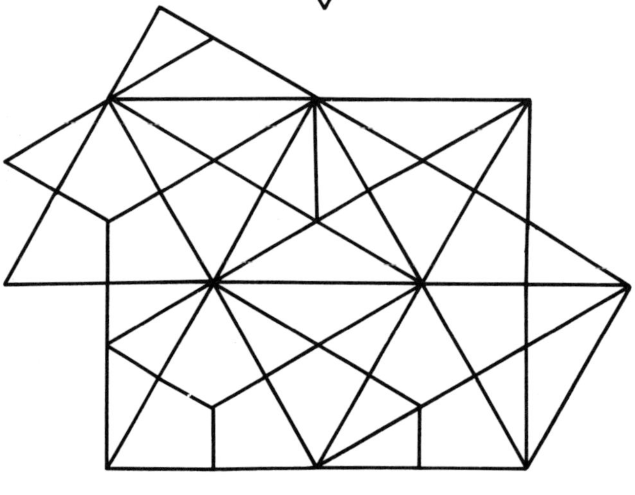

Dual Tessellation

A dual tessellation of a tiling is formed by joining the center of each polygon through its sides to the centers of all neighboring polygons. A dual tessellation has as many polygons as the original tessellation has vertices, and as many vertices as the original has polygons. The number of sides remains the same.

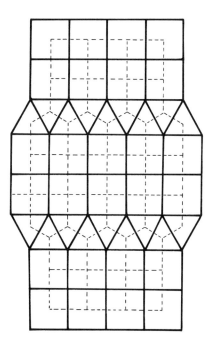

A polygon formed by a dual tessellation has the same number of sides as there are edges meeting in its center.

The two regular tessellations of triangles and hexagons are dual to each other.

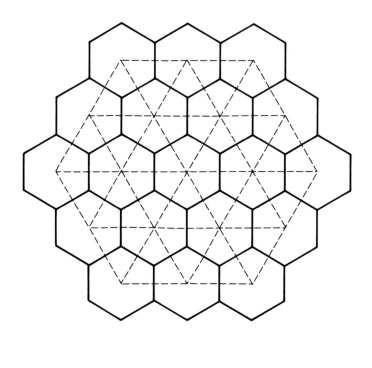

Self-Dual

A self-dual forms the same polygons as comprise the original tessellation. The regular tessellation of squares is a self-dual.

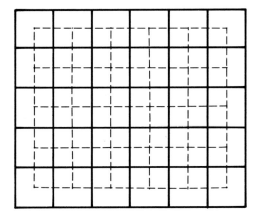

Dual Tilings of the Semiregular Tessellations

The dual tilings of the semiregular tessellations form polygons that are all nonregular. Because the semiregular tessellations are uniform, the polygons formed by the dual tiling are congruent or enantiomorphic. The dual tilings are periodic but not uniform.

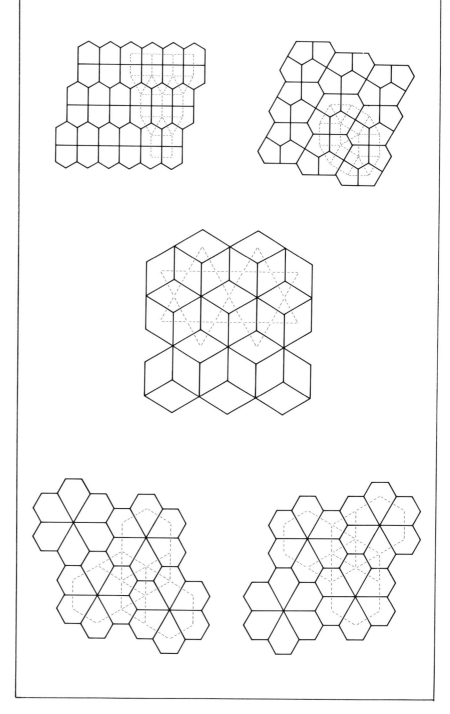

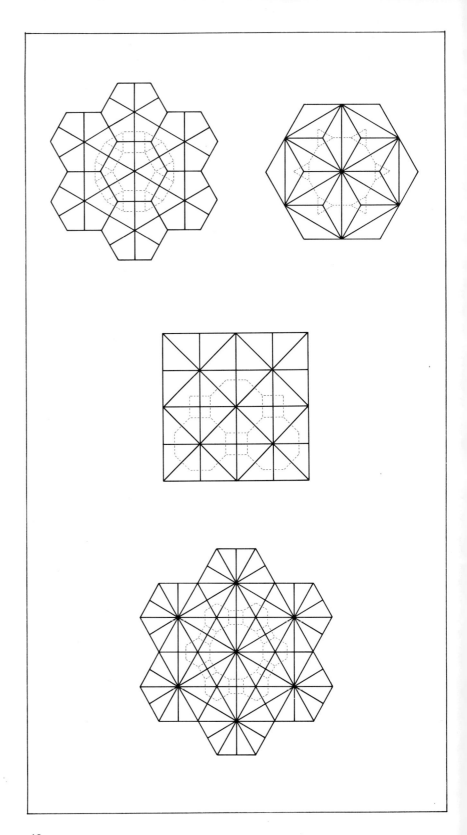

Tessellations and Symmetry

In a tiling with congruent polygons, only polygons with the following symmetries may be used: no symmetry, mirror symmetry, 2-fold, 3-fold, 4-fold, and 6-fold. Polygons with other symmetries, such as 5-fold, will not tile the plane.

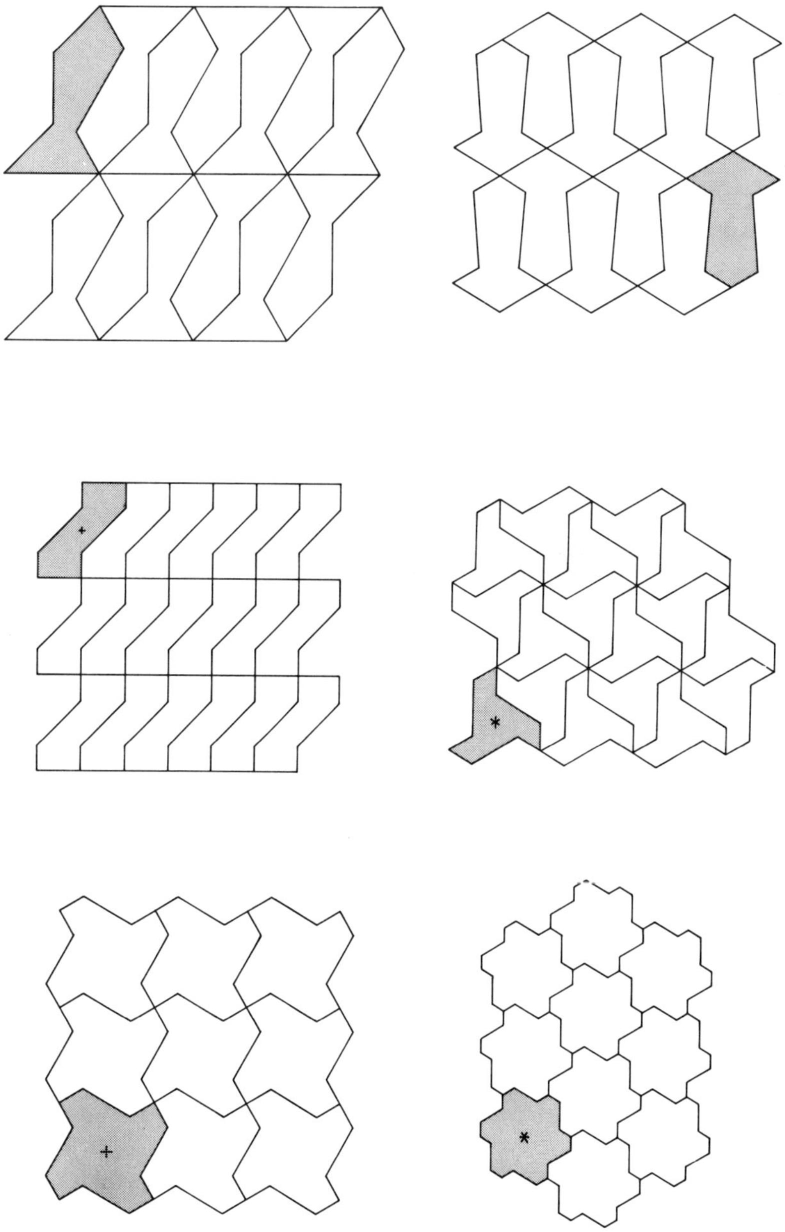

Open Patterns with Regular Polygons

Periodic Patterns

The plane can be subdivided periodically without the requirement of filling the whole plane. In such cases, regular polygons of 5, 9, 10, and 20 sides may be used. Use of these polygons results in open spaces of nonregular polygons.

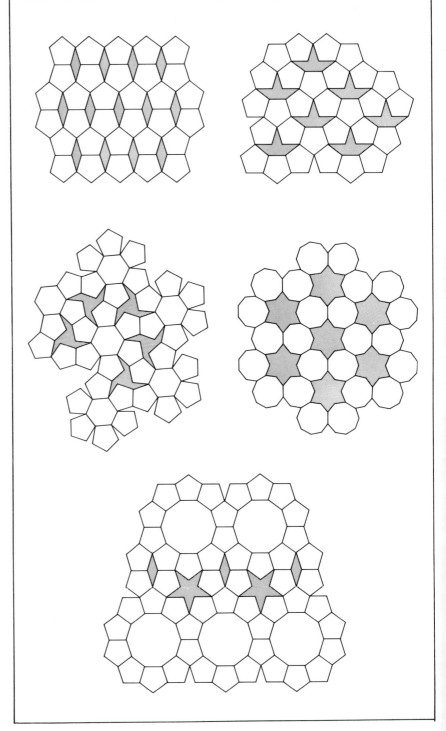

Concentric Patterns

Though the regular pentagon does not tile the plane, it can be used for open patterns that form concentrically around a central figure.

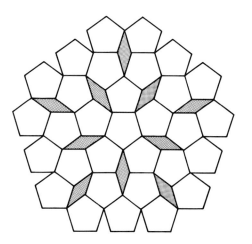

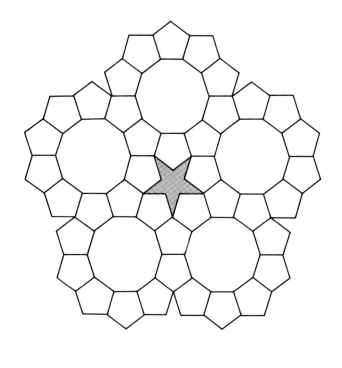

Euler's Theorem

Euler's theorem for tilings demonstrates that there is always a consistent relationship among the components of a tessellation: the number of polygons + the number of vertices = the number of edges + 1.

$$P + V = E + 1$$

$$6 + 7 = 12 + 1$$
$$13 = 13$$

$$4 + 10 = 13 + 1$$
$$14 = 14$$

3 Polyhedra

Polyhedron

A polyhedron is formed by enclosing a portion of three-dimensional space with four or more plane polygons. (Plural: polyhedra).

Faces

The faces of a polyhedron are polygons.

Edges

The edges of a polyhedron are formed where common sides of neighboring polygons meet.

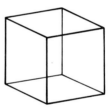

Vertex

A vertex of a polyhedron is a point where edges intersect. Three or more polygons must meet at each vertex. The sum of the face angles of the polygons meeting at a vertex <mark>must always equal less than 360°</mark>

Dihedral Angle

A dihedral angle is the angle formed by two polygons joined along a common edge.

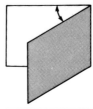

Convex Polyhedron | A polyhedron is convex if every dihedral angle is less than 180°.

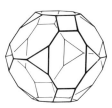

Concave Polyhedron | A polygon is concave if at least one of its dihedral angles is more than 180°.

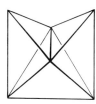

Uniform | A polyhedron is uniform if all of its vertices are the same or congruent.

Euler's Theorem for Polyhedra | Euler's theorem for polyhedra demonstrates the constant relationship among the components of a given polyhedron: the number of polygons + the number of vertices = the number of edges + 2.

$$P + V = E + 2$$

$$4 + 4 = 6 + 2$$
$$8 = 8$$

$$6 + 8 = 12 + 2$$
$$14 = 14$$

Naming Polyhedra

n-hedron

Polyhedra are usually named by the number of faces they have. An n-hedron is a polyhedron with an unspecified number of faces.

A *tetra*hedron has 4 faces.

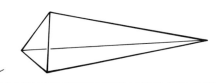

A *penta*hedron has 5 faces.

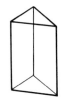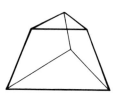

A *hexa*hedron has 6 faces.

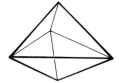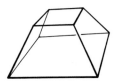

A *septa*hedron has 7 faces.

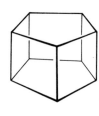

An *octa*hedron has 8 faces.

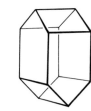

A *deca*hedron has 10 faces.

A *dodeca*hedron has 12 faces.

  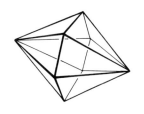

A *tetrakaideca*hedron has 14 faces.

A *pentakaideca*hedron has 15 faces.

A *hexakaideca*hedron has 16 faces.

 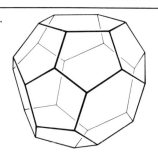

An *icosa*hedron has 20 faces.

An *icosiocta*hedron has 28 faces.

An *icosidodeca*hedron has 32 faces.

A *pentaconta*hedron has 50 faces.

Polyhedra and Symmetry

The symmetry properties of a polyhedron are determined by viewing it from different orientations. The number of orientations that produce views of symmetry may vary for different polyhedra. Views of symmetry are characterized as rotational and/or mirror. A particular view of symmetry may occur more than once in a given polyhedron. One possible view of symmetry may be determined by looking at a polyhedron toward a centered vertex.

Another view of symmetry may be determined by looking at a polyhedron toward a centered edge.

A third view of symmetry may be determined by looking at a polyhedron toward a centered face.

Regular Polyhedra

Regular polyhedra are uniform and have faces of all of one kind of congruent regular polygon. There are five regular polyhedra. The regular polyhedra were an important part of Plato's natural philosophy, and thus have come to be called the Platonic Solids.

In the views of symmetry shown below, each type of rotational symmetry and the number of times it occurs is indicated by: n-fold(x). The mirror planes are represented by dotted lines.

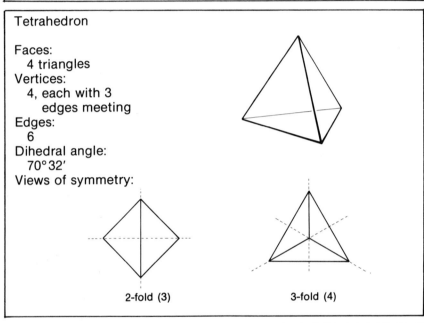

Tetrahedron

Faces:
　4 triangles
Vertices:
　4, each with 3 edges meeting
Edges:
　6
Dihedral angle:
　70° 32′
Views of symmetry:

2-fold (3)　　　3-fold (4)

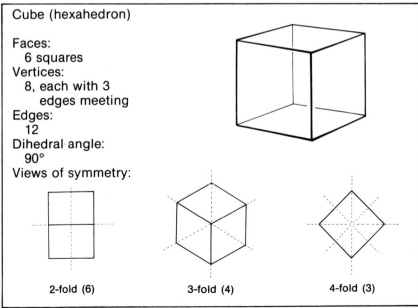

Cube (hexahedron)

Faces:
　6 squares
Vertices:
　8, each with 3 edges meeting
Edges:
　12
Dihedral angle:
　90°
Views of symmetry:

2-fold (6)　　　3-fold (4)　　　4-fold (3)

Octahedron

Faces:
 8 triangles
Vertices:
 6, each with 4
  edges meeting
Edges:
 12
Dihedral angle:
 109° 28'
Views of symmetry:

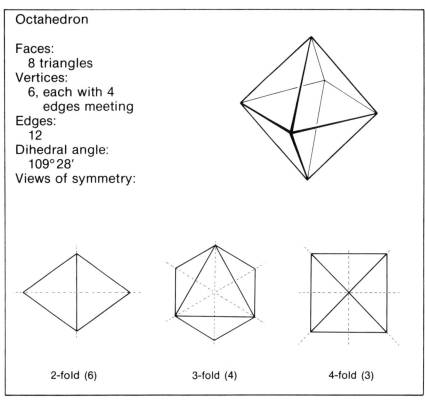

2-fold (6) 3-fold (4) 4-fold (3)

Dodecahedron

Faces:
 12 pentagons
Vertices:
 20, each with 3
  edges meeting
Edges:
 30
Dihedral angle:
 116° 34'
Views of symmetry:

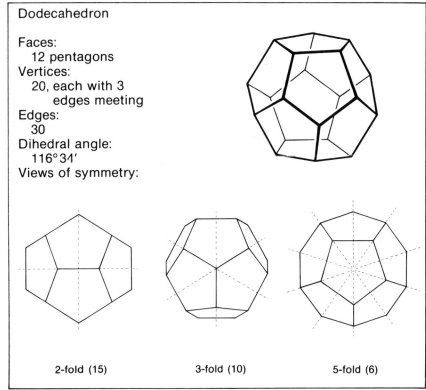

2-fold (15) 3-fold (10) 5-fold (6)

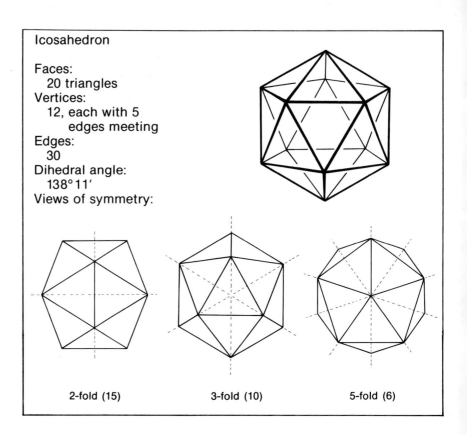

Icosahedron

Faces:
  20 triangles
Vertices:
  12, each with 5 edges meeting
Edges:
  30
Dihedral angle:
  138° 11'
Views of symmetry:

2-fold (15)   3-fold (10)   5-fold (6)

Semiregular Polyhedra

A semiregular polyhedron has regular polygons as faces, but the faces are not all of the same kind. As in regular polyhedra, the vertices are congruent. There are thirteen semiregular polyhedra. It is generally believed that they were described by Archimedes, and thus are called the Archimedean Polyhedra.

As in the regular polyhedra, in the views of symmetry shown below, each type of rotational symmetry and the number of times it occurs is indicated by: n-fold(x), and the mirror planes are represented by dotted lines.

Five semiregular polyhedra are derived by truncating the five regular polyhedra. Truncation is done so that all new faces are regular polygons. The polyhedra formed are the truncated tetrahedron, truncated cube, truncated octahedron, truncated dodecahedron, and truncated icosahedron.

Truncated Tetrahedron

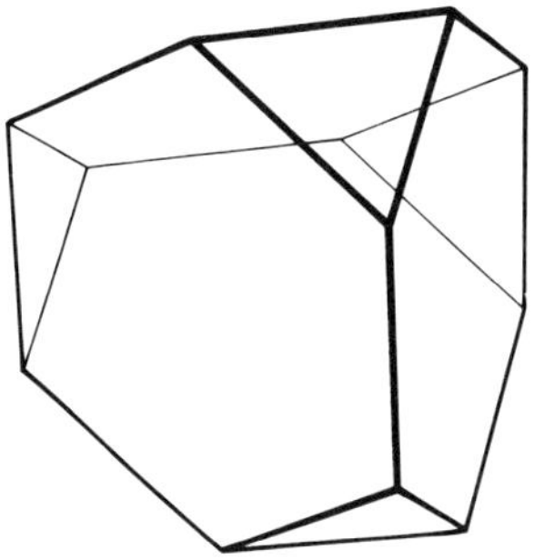

Faces:
  4 hexagons ⎤ 8 total
  4 triangles ⎦
Vertices:
  12, each with 3 edges meeting
Edges:
  18
Dihedral angles:
  70°32' (hexagon-hexagon)
  109°28' (triangle-hexagon)
Views of symmetry:

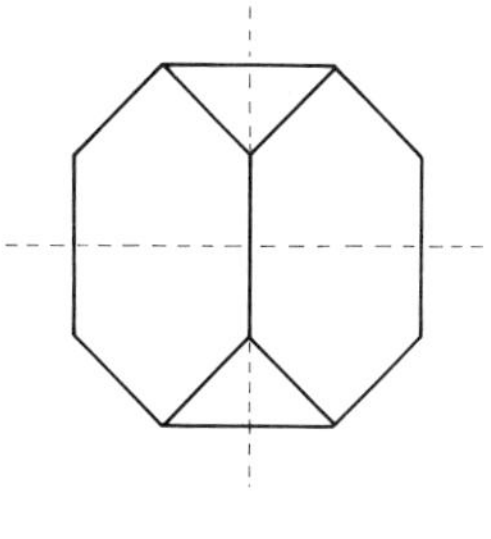

2-fold (3) 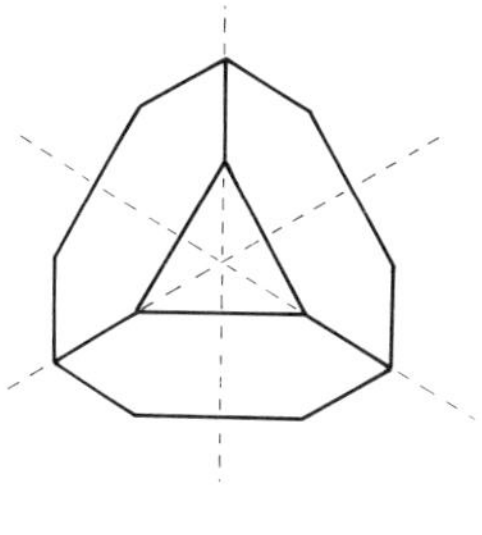   3-fold (4)

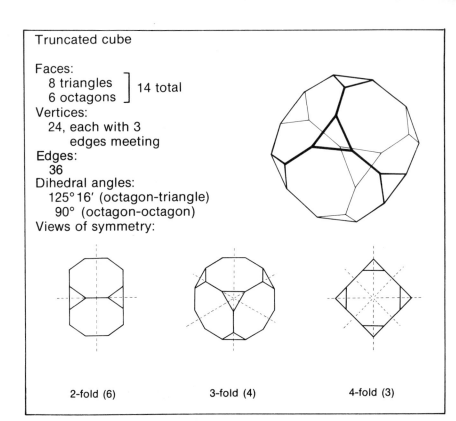

Truncated cube

Faces:
 8 triangles ⎤ 14 total
 6 octagons ⎦
Vertices:
 24, each with 3 edges meeting
Edges:
 36
Dihedral angles:
 125° 16′ (octagon-triangle)
 90° (octagon-octagon)
Views of symmetry:

2-fold (6)   3-fold (4)   4-fold (3)

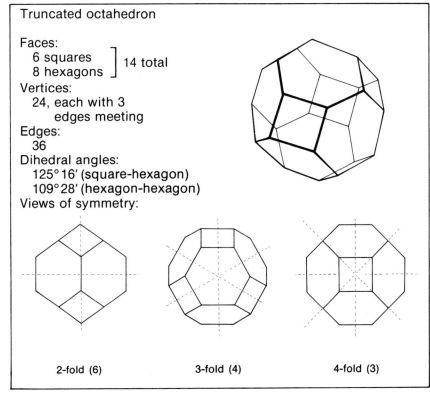

Truncated octahedron

Faces:
 6 squares ⎤ 14 total
 8 hexagons ⎦
Vertices:
 24, each with 3 edges meeting
Edges:
 36
Dihedral angles:
 125° 16′ (square-hexagon)
 109° 28′ (hexagon-hexagon)
Views of symmetry:

2-fold (6)   3-fold (4)   4-fold (3)

Truncated dodecahedron

Faces:
  20 triangles ⎱ 32 total
  12 decagons ⎰
Vertices:
  60, each with 3 edges meeting
Edges:
  90
Dihedral angles:
  116°34' (decagon-decagon)
  142°37' (decagon-triangle)
Views of symmetry:

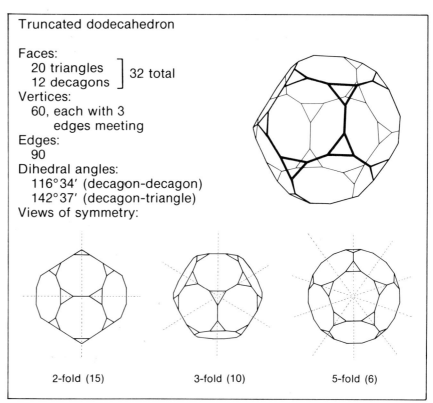

2-fold (15)   3-fold (10)   5-fold (6)

Truncated icosahedron

Faces:
  12 pentagons ⎱ 32 total
  20 hexagons ⎰
Vertices:
  60, each with 3 edges meeting
Edges:
  90
Dihedral angles:
  138°11' (hexagon-hexagon)
  142°37' (hexagon-pentagon)
Views of symmetry:

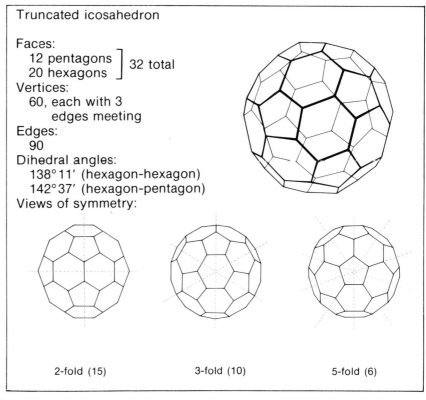

2-fold (15)   3-fold (10)   5-fold (6)

Quasiregular Polyhedra

A quasiregular polyhedron has two kinds of faces, with each face of one kind being entirely surrounded by the face of the other kind. Two semiregular polyhedra are quasiregular: the cuboctahedron and the icosidodecahedron.

Cuboctahedron

Faces:
 8 triangles ⎱ 14 total
 6 squares ⎰
Vertices:
 12, each with 4 edges meeting
Edges:
 24
Dihedral angle:
 125° 16′
Views of symmetry:

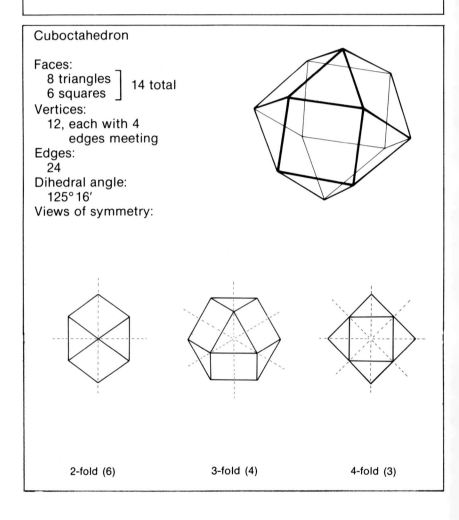

2-fold (6)   3-fold (4)   4-fold (3)

Icosidodecahedron

Faces:
  20 triangles ⎤
  12 pentagons ⎦ 32 total
Vertices:
  30, each with 4
      edges meeting
Edges:
  60
Dihedral angle:
  142°37′
Views of symmetry:

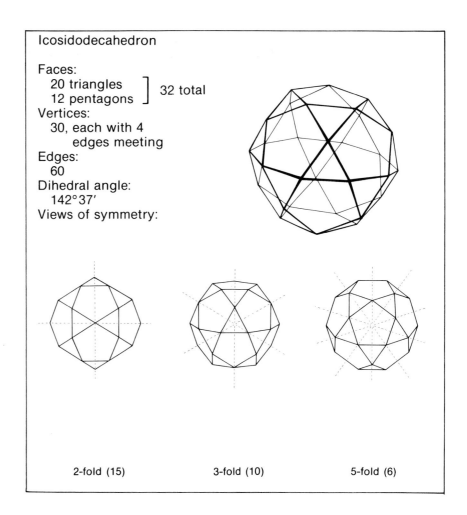

2-fold (15)   3-fold (10)   5-fold (6)

The snub cuboctahedron (also known as the snub cube) and the snub icosidodecahedron (also known as the snub dodecahedron) are derivations of the cuboctahedron and icosidodecahedron respectively that are formed by adding extra equilateral triangles to the original polyhedron.

Snub cuboctahedron

Faces:
  32 triangles ⎤ 38 total
  6 squares ⎦
Vertices:
  24, each with 5 edges meeting
Edges:
  60
Dihedral angles:
  142°59′ (square-triangle)
  153°14′ (triangle-triangle)
Views of symmetry:

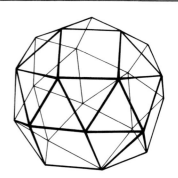

2-fold (6)   3-fold (4)   4-fold (3)

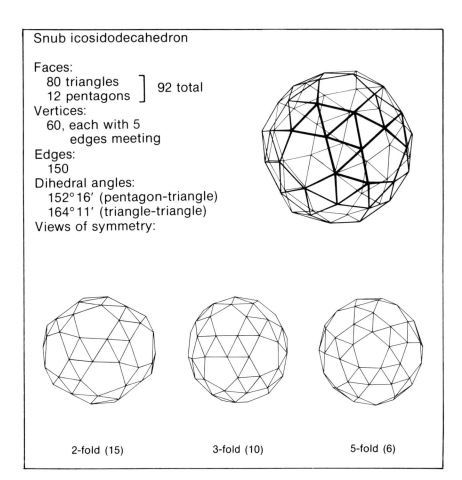

Snub icosidodecahedron

Faces:
   80 triangles ⎤ 92 total
   12 pentagons ⎦
Vertices:
   60, each with 5
      edges meeting
Edges:
   150
Dihedral angles:
   152°16′ (pentagon-triangle)
   164°11′ (triangle-triangle)
Views of symmetry:

2-fold (15)   3-fold (10)   5-fold (6)

Truncating the cuboctahedron in two different ways gives rise to the truncated cuboctahedron (also known as the greater rhombicuboctahedron) and the rhombicuboctahedron.

Truncated cuboctahedron
(Greater rhombicuboctahedron)

Faces:
  12 squares ⎤
  8 hexagons ⎬ 26 total
  6 octagons ⎦
Vertices:
  48, each with 3 edges meeting
Edges:
  72
Dihedral angles:
  135° (octagon-square)
  125° 16′ (octagon-hexagon)
  144° 44′ (hexagon-square)
Views of symmetry:

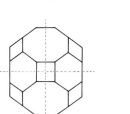
2-fold (6)

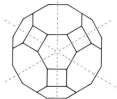
3-fold (4)

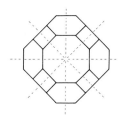
4-fold (3)

Rhombicuboctahedron

Faces:
  8 triangles ⎤ 26 total
  18 squares ⎦
Vertices:
  24, each with 4 edges meeting
Edges:
  48
Dihedral angles:
  135° (square-square)
  144°44′ (square-triangle)
Views of symmetry:

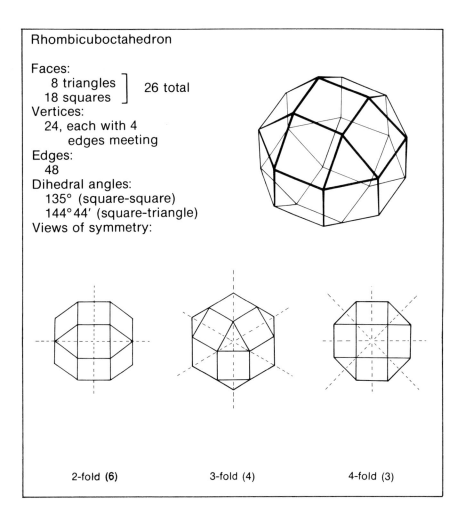

2-fold **(6)**   3-fold **(4)**   4-fold **(3)**

Truncating the icosidodecahedron in two different ways gives rise to the truncated icosidodecahedron (also known as the greater rhombicosidodecahedron), and the rhombicosidodecahedron.

Truncated icosidodecahedron
(Greater rhombicosidodecahedron)

Faces:
   30 squares ⎤
   20 hexagons ⎬ 62 total
   12 decagons ⎦
Vertices:
   120, each with 3
         edges meeting
Edges:
   180
Dihedral angles:
   148° 17' (decagon-square)
   142° 37' (decagon-hexagon)
   159° 6' (hexagon-square)
Views of symmetry:

2-fold (15)   3-fold (10)   5-fold (6)

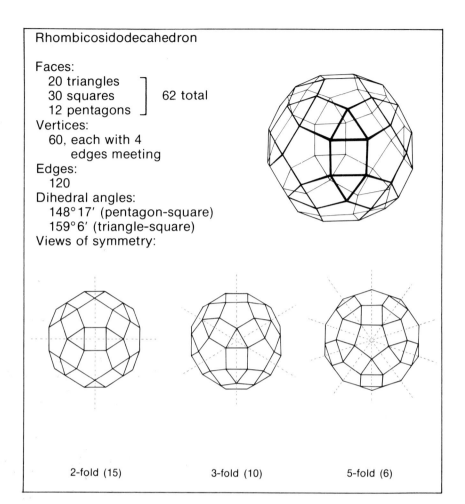

Rhombicosidodecahedron

Faces:
   20 triangles ⎤
   30 squares   ⎬ 62 total
   12 pentagons ⎦
Vertices:
   60, each with 4
      edges meeting
Edges:
   120
Dihedral angles:
   148° 17′ (pentagon-square)
   159° 6′ (triangle-square)
Views of symmetry:

2-fold (15)   3-fold (10)   5-fold (6)

Prism

A prism is a polyhedron with two congruent and parallel faces that are joined by a set of parallelograms. The prism is semi-regular if all the polygons are regular. A cube can be considered a square prism.

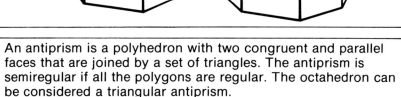

Antiprism

An antiprism is a polyhedron with two congruent and parallel faces that are joined by a set of triangles. The antiprism is semiregular if all the polygons are regular. The octahedron can be considered a triangular antiprism.

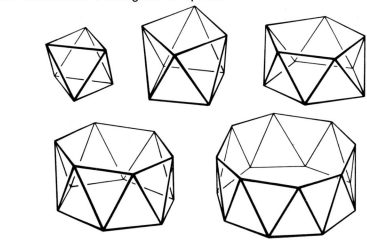

Prisms and antiprisms correspond to the infinite number of possible polygons.

Deltahedra

Deltahedra are convex polyhedra whose faces consist entirely of equilateral triangles. The deltahedra are the only polyhedra, beside the regular polyhedra, that have faces consisting of only one kind of regular polygon. Three of the eight deltahedra are regular polyhedra. The remaining five do not have congruent vertices.

Tetrahedron (4-hedron)

Triangular Dipyramid (6-hedron)

Octahedron (8-hedron)

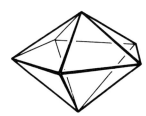

Pentagonal Dipyramid (10-hedron)

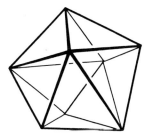

12-hedron

14-hedron

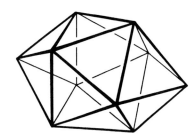

16-hedron

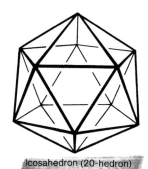

Icosahedron (20-hedron)

Other Convex Polyhedra with Regular Faces

In addition to the regular and semiregular polyhedra, the deltahedra, and the infinite class of prisms and antiprisms, there are 87 more convex polyhedra whose faces are entirely regular polygons. These polyhedra do not have congruent vertices. Only regular triangles, squares, pentagons, hexagons, octagons, and decagons may be used for the formation of polyhedra with regular faces. Some of these additional polyhedra are shown below.

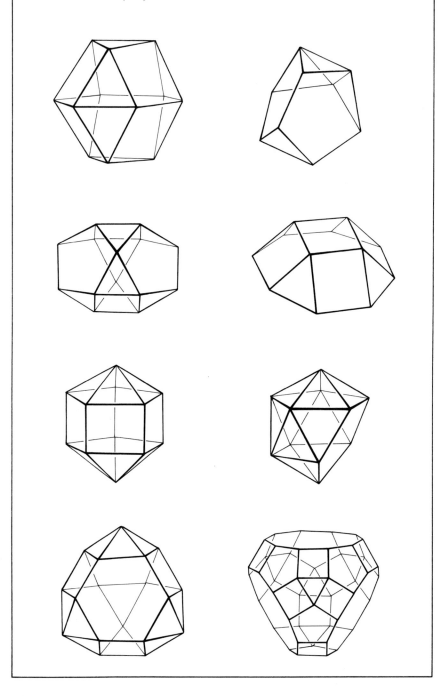

Stellated Polyhedra

A stellated polyhedron is formed by extending in the same plane each face of a convex polyhedron until the faces intersect to form a new enclosing shape. Usually stellations are performed on polyhedra whose faces are all alike, but the operation can be done on other polyhedra.

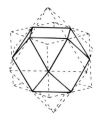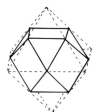

Stella Octangula

The stella octangula is the polyhedron formed by stellating an octahedron. The stella octangula can also be thought of as an intersection of two tetrahedra.

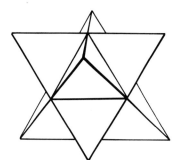

Kepler-Poinsot Solids

The stellations of the dodecahedron and icosahedron are called the Kepler-Poinsot Solids. The regular dodecahedron can be stellated to form the small stellated dodecahedron.

The regular dodecahedron can be stellated in a different manner to form the great dodecahedron.

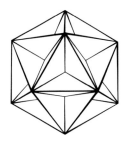

The dodecahedron can again be stellated to form the great stellated dodecahedron.

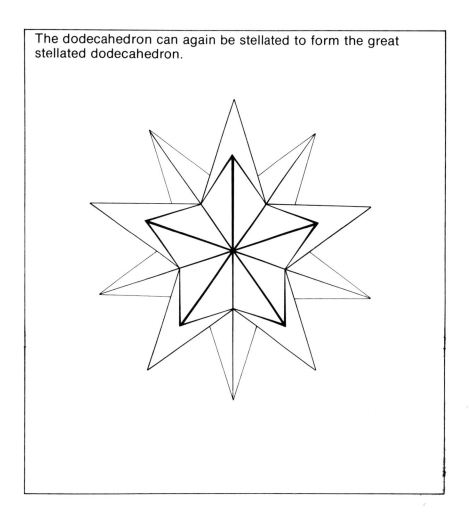

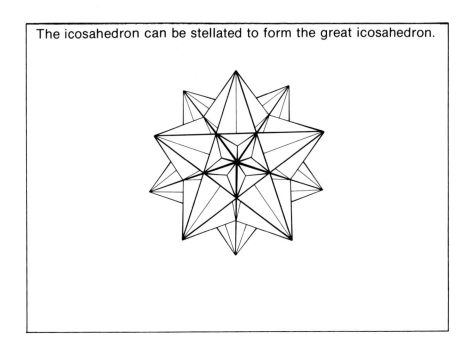
The icosahedron can be stellated to form the great icosahedron.

4　Dual Polyhedra

Dual Polyhedra

A dual polyhedron is formed by joining a point that is perpendicularly above the center of each face of a polyhedron to equivalent points above all neighboring faces. The new edges connecting these points intersect the edges of the original polyhedron. A polyhedron and its dual each have the same number of edges. A dual polyhedron has as many vertices as the original polyhedron has faces, and the dual has as many faces as the original has vertices. The symmetry of a dual polyhedron is the same as that of the original polyhedron.

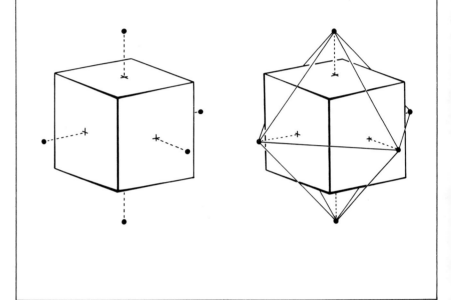

A pair of dual polyhedra must enclose a common sphere that is tangent to both at the points where their respective edges meet.

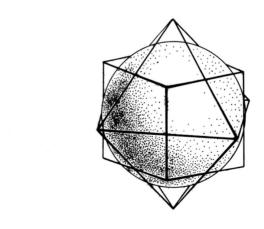

Duals of the Regular Polyhedra

The dual of a regular tetrahedron is another regular tetrahedron. It is a self-dual.

Tetrahedron:
  4 faces, 4 vertices, 6 edges

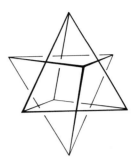

The cube and the octahedron are dual to each other.

Cube:
  6 faces, 8 vertices, 12 edges
Octahedron:
  8 faces, 6 vertices, 12 edges

The dodecahedron and the icosahedron are dual to each other.

Dodecahedron:
  12 faces, 20 vertices, 30 edges
Icosahedron:
  20 faces, 12 vertices, 30 edges

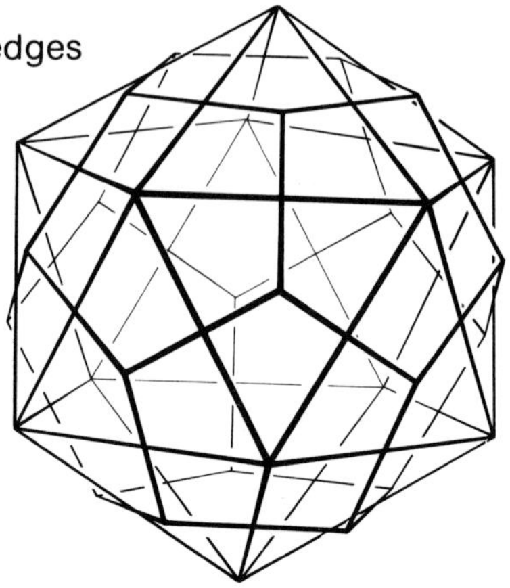

Duals of the Semiregular Polyhedra

Each dual of the semiregular polyhedra has congruent faces, but none of the faces are regular polygons. The faces of the dual are congruent because the vertices of the original polyhedron are congruent. The dual polyhedra of the semiregular polyhedra are analogous to the dual networks of the semiregular tessellations.

Again, in the views of symmetry shown below, each type of rotational symmetry and the number of times it occurs is indicated by: n-fold(x), and the mirror planes are represented by dotted lines.

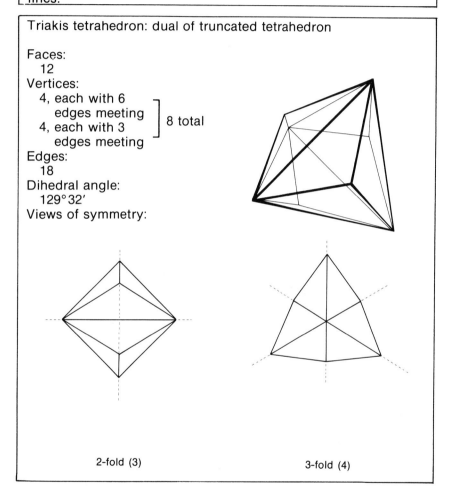

Triakis tetrahedron: dual of truncated tetrahedron

Faces:
  12
Vertices:
  4, each with 6 edges meeting ⎤
  4, each with 3 edges meeting ⎦ 8 total
Edges:
  18
Dihedral angle:
  129° 32'
Views of symmetry:

2-fold (3)   3-fold (4)

Triakis octahedron: dual of truncated cube

Faces:
 24
Vertices:
 8, each with 3 edges meeting ⎤
 6, each with 8 edges meeting ⎦ 14 total
Edges:
 36
Dihedral angle:
 147° 21′
Views of symmetry:

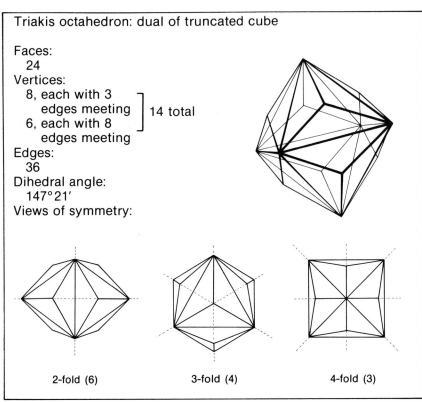

2-fold (6)   3-fold (4)   4-fold (3)

Tetrakis hexahedron: dual of truncated octahedron

Faces:
 24
Vertices:
 6, each with 4 edges meeting ⎤
 8, each with 6 edges meeting ⎦ 14 total
Edges:
 36
Dihedral angle:
 143° 8′
Views of symmetry:

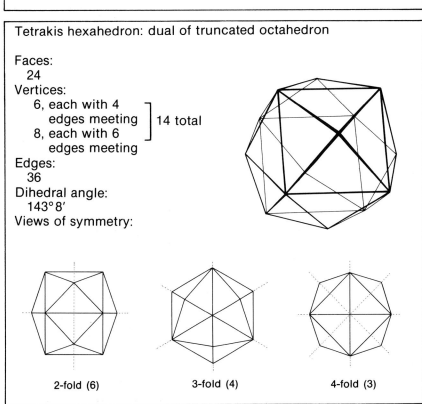

2-fold (6)   3-fold (4)   4-fold (3)

Triakis icosahedron: dual of truncated dodecahedron

Faces:
  60
Vertices:
  20, each with 3 edges meeting
  12, each with 10 edges meeting
  } 32 total
Edges:
  90
Dihedral angle:
  160° 36′
Views of symmetry:

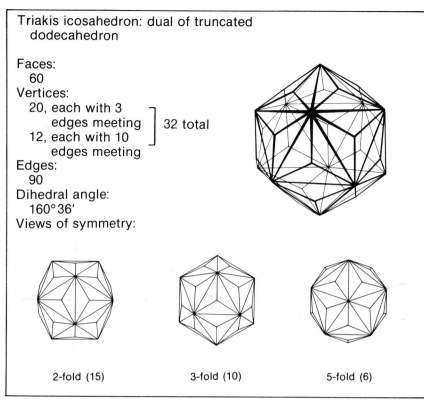

2-fold (15) 3-fold (10) 5-fold (6)

Pentakis dodecahedron: dual of truncated icosahedron

Faces:
  60
Vertices:
  12, each with 5 edges meeting
  20, each with 6 edges meeting
  } 32 total
Edges:
  90
Dihedral angle:
  156° 43′
Views of symmetry:

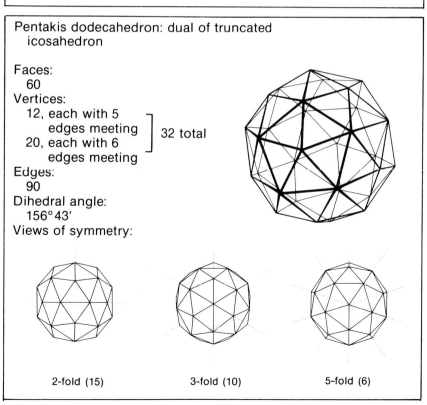

2-fold (15) 3-fold (10) 5-fold (6)

Rhombic dodecahedron: dual of cuboctahedron

Faces:
  12
Vertices:
  8, each with 3 edges meeting ⎤
  6, each with 4 edges meeting ⎦ 14 total
Edges:
  24
Dihedral angle:
  120°
Views of symmetry:

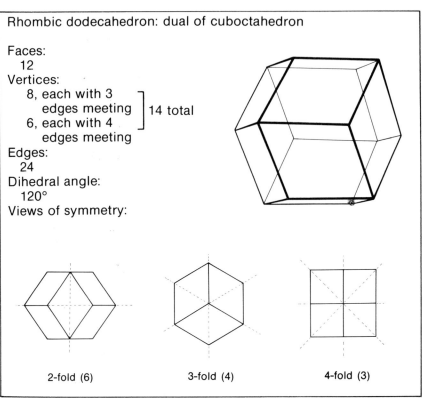

2-fold (6)   3-fold (4)   4-fold (3)

Rhombic triacontahedron: dual of icosido-decahedron

Faces:
  30
Vertices:
  20, each with 3 edges meeting ⎤
  12, each with 5 edges meeting ⎦ 32 total
Edges:
  60
Dihedral angle:
  144°
Views of symmetry:

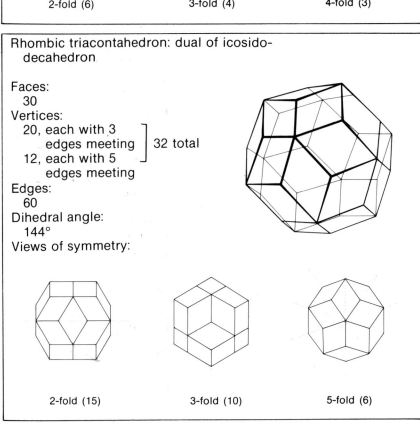

2-fold (15)   3-fold (10)   5-fold (6)

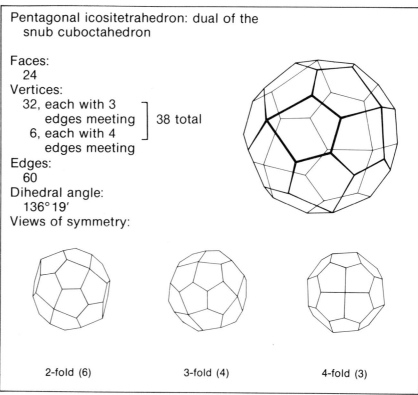

Pentagonal icositetrahedron: dual of the snub cuboctahedron

Faces:
  24
Vertices:
  32, each with 3 edges meeting ⎤ 38 total
  6, each with 4 edges meeting ⎦
Edges:
  60
Dihedral angle:
  136° 19′
Views of symmetry:

2-fold (6)   3-fold (4)   4-fold (3)

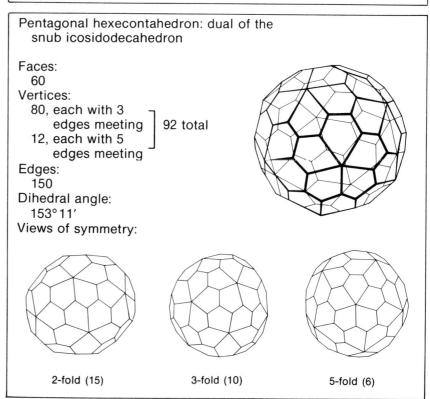

Pentagonal hexecontahedron: dual of the snub icosidodecahedron

Faces:
  60
Vertices:
  80, each with 3 edges meeting ⎤ 92 total
  12, each with 5 edges meeting ⎦
Edges:
  150
Dihedral angle:
  153° 11′
Views of symmetry:

2-fold (15)   3-fold (10)   5-fold (6)

Hexakis octahedron: dual of truncated cuboctahedron

Faces:
  48
Vertices:
  12, each with 4 edges meeting
  8, each with 6 edges meeting   } 26 total
  6, each with 8 edges meeting
Edges:
  72
Dihedral angle:
  155° 5′
Views of symmetry:

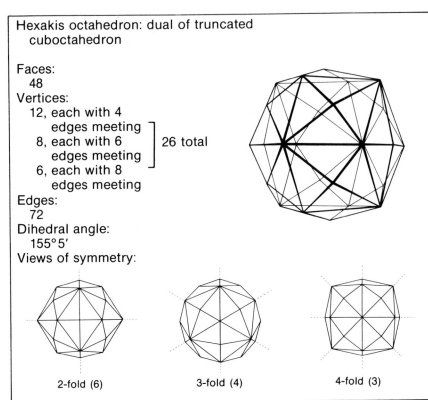

2-fold (6)   3-fold (4)   4-fold (3)

Trapezoidal icositetrahedron: dual of rhombicuboctahedron

Faces:
  24
Vertices:
  8, each with 3 edges meeting   } 26 total
  18, each with 4 edges meeting
Edges:
  48
Dihedral angle
  138° 7′
Views of symmetry:

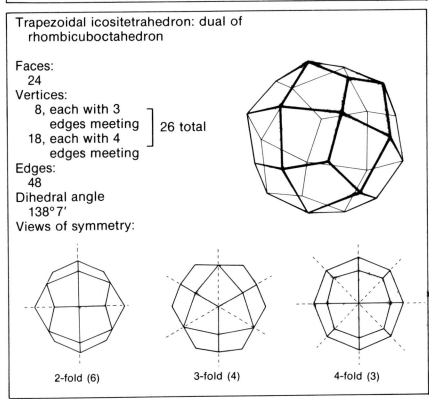

2-fold (6)   3-fold (4)   4-fold (3)

Hexakis icosahedron: dual of truncated icosidodecahedron

Faces:
 120
Vertices:
 30, each with 4 edges meeting ⎤
 20, each with 6 edges meeting ⎬ 62 total
 12, each with 10 edges meeting ⎦
Edges:
 150
Dihedral angle:
 164° 54'
Views of symmetry:

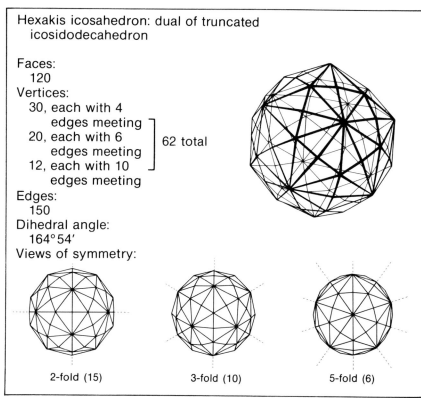

2-fold (15)   3-fold (10)   5-fold (6)

Trapezoidal hexecontahedron: dual of rhombicosidodecahedron

Faces:
 60
Vertices:
 20, each with 3 edges meeting ⎤
 30, each with 4 edges meeting ⎬ 62 total
 12, each with 5 edges meeting ⎦
Edges:
 120
Dihedral angle:
 154° 8'
Views of symmetry:

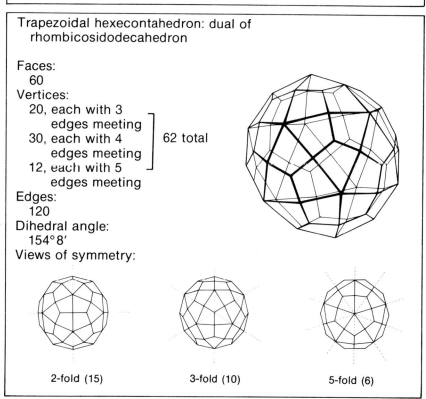

2-fold (15)   3-fold (10)   5-fold (6)

Duals of the Prisms

The duals of the prisms are dipyramids and have faces that are congruent isosceles triangles.

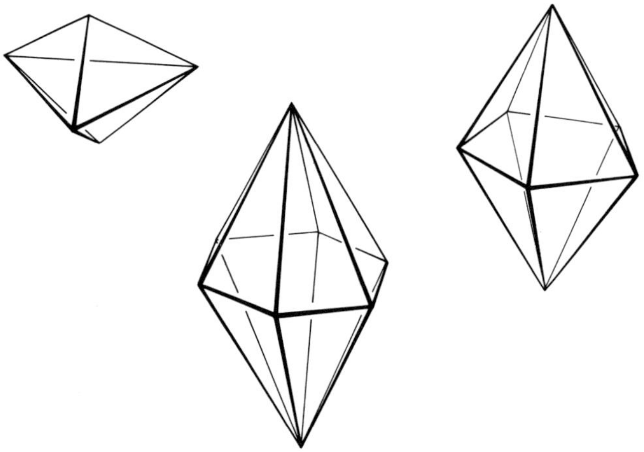

Duals of the Antiprisms

The duals of the antiprisms are trapezohedra and have faces that are congruent trapezia.

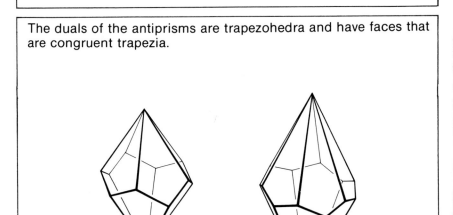

5 Space Filling

Combining Polyhedra

Polyhedra can be combined by joining them at their matching or congruent faces.

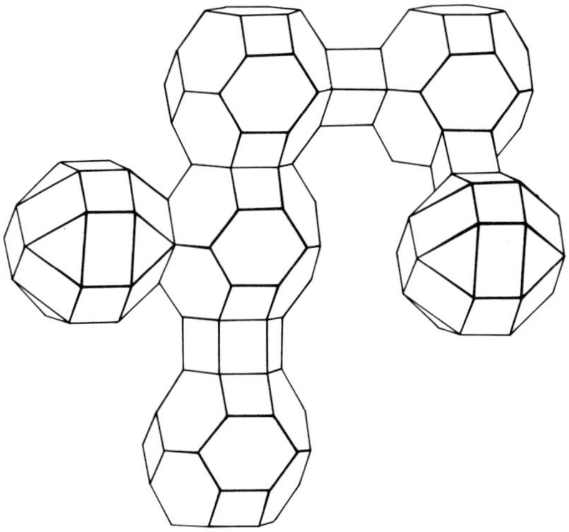

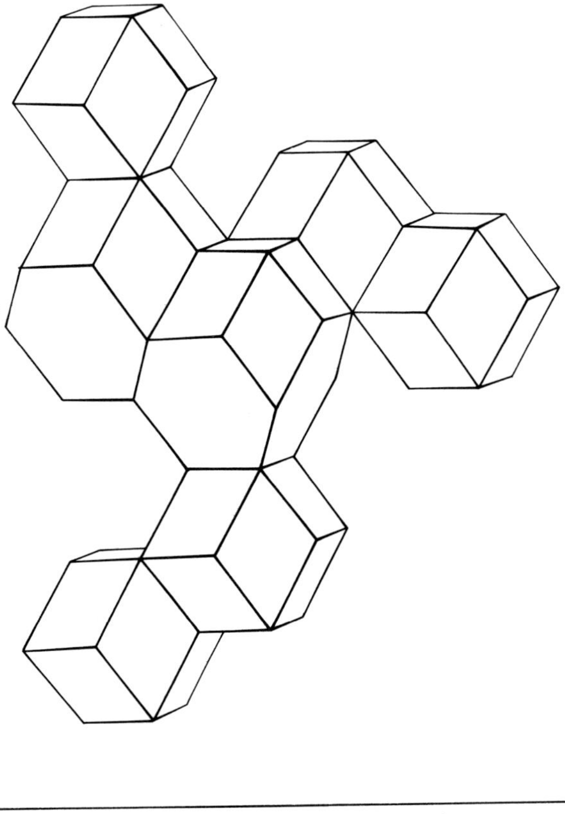

Space Filling

Space filling occurs when polyhedra are packed together in a repeating array so that all volume is occupied and there is no left-over space.

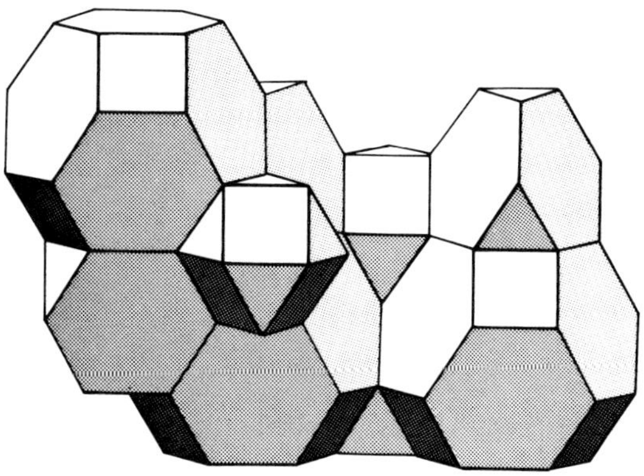

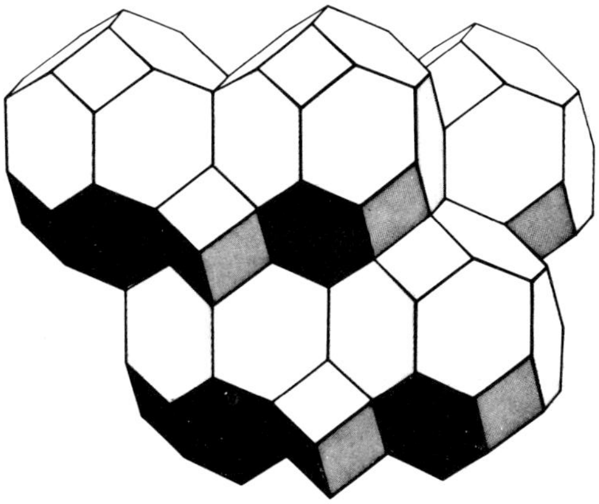

Dihedral Angles and Space Filling

In order for polyhedra to fill space with each other, the sum of those dihedral angles that occur around a common edge must equal 360°. This is analogous to the 360° requirement around a single point in a plane tessellation. The dihedral angle of a tetrahedron is 70°32′ so it will not fill space alone because there is no multiple of its dihedral angle that equals 360°. It will fill space, however, with octahedra whose dihedral angles are 109°28′. 109°28′ + 70°32′ + 109°28′ + 70°32′ = 360°.

The dihedral angle of the cube is 90°, and since 90° × 4 = 360°, the cube will fill space.

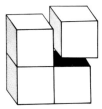

Network

A network is formed by the edges of the polyhedra in a space filling array.

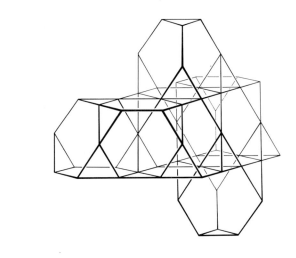

Uniform Network

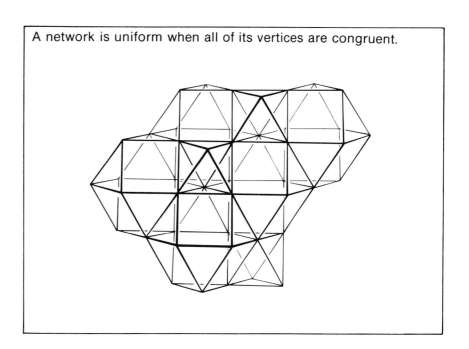

A network is uniform when all of its vertices are congruent.

Uniform Space Filling with One Kind of Polyhedron

Regular

Of the regular polyhedra, only the cube will fill space by itself.

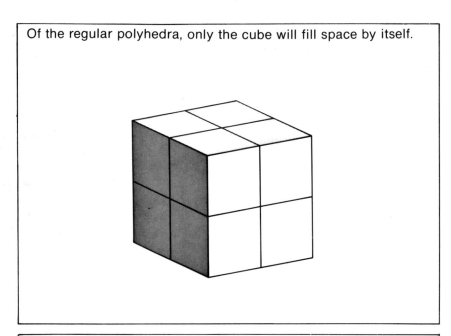

Semiregular

Of the semiregular polyhedra, only the truncated octahedron will fill space by itself.

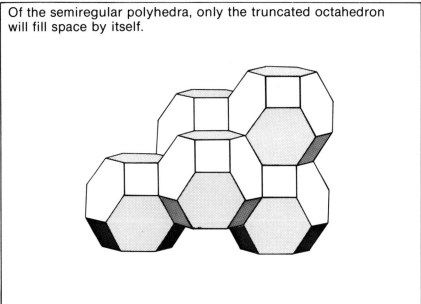

Prisms

The triangular prism will fill space.

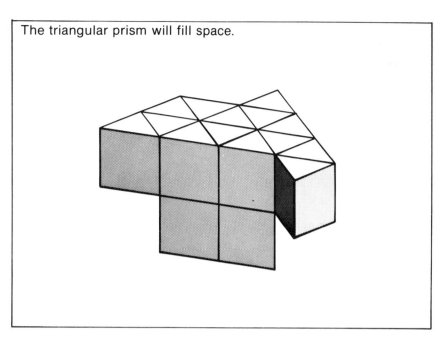

The hexagonal prism will fill space.

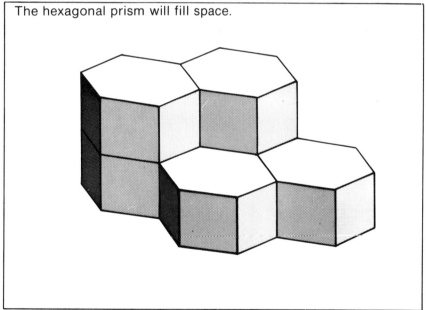

Uniform Space Filling with Two Kinds of Polyhedra

Regular

Tetrahedra/octahedra. The relative numbers of each polyhedron that will combine to fill space is expressed as a space filling ratio. In this case, the ratio of tetrahedra to octahedra is 2:1.

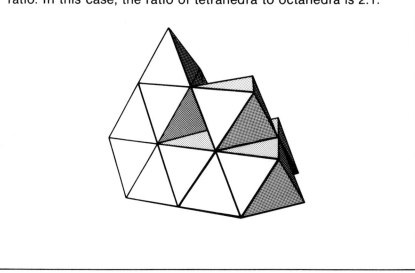

Regular/ Semiregular

Tetrahedra/truncated tetrahedra. Space filling ratio — 1:1

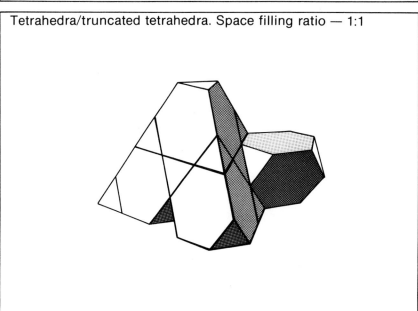

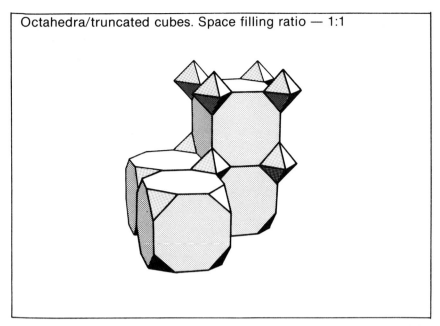

Octahedra/truncated cubes. Space filling ratio — 1:1

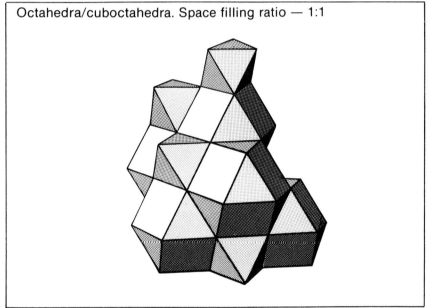

Octahedra/cuboctahedra. Space filling ratio — 1:1

Semiregular/ Prisms

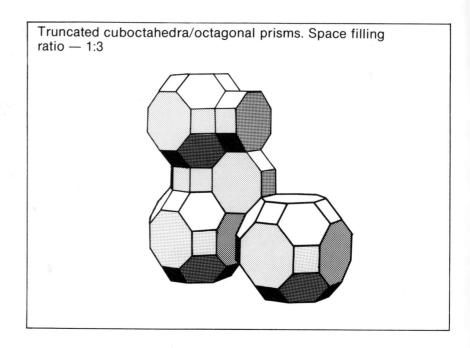

Truncated cuboctahedra/octagonal prisms. Space filling ratio — 1:3

Uniform Space Filling with Three Kinds of Polyhedra

Regular/Semiregular

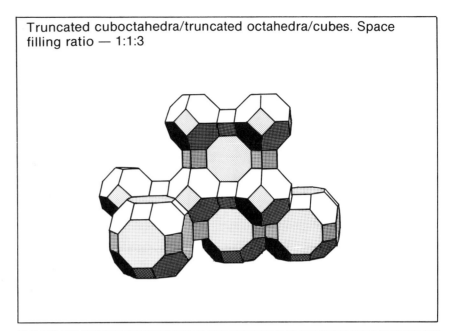

Truncated cuboctahedra/truncated octahedra/cubes. Space filling ratio — 1:1:3

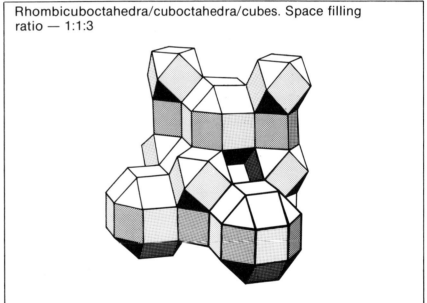

Rhombicuboctahedra/cuboctahedra/cubes. Space filling ratio — 1:1:3

Semiregular

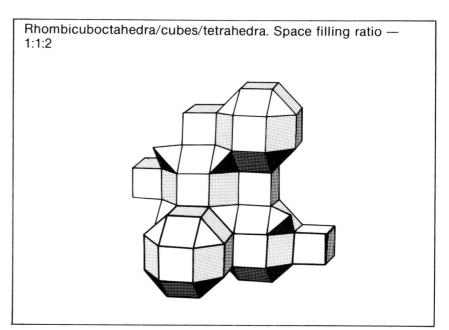

Rhombicuboctahedra/cubes/tetrahedra. Space filling ratio — 1:1:2

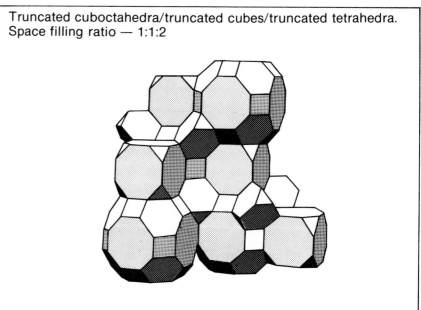

Truncated cuboctahedra/truncated cubes/truncated tetrahedra. Space filling ratio — 1:1:2

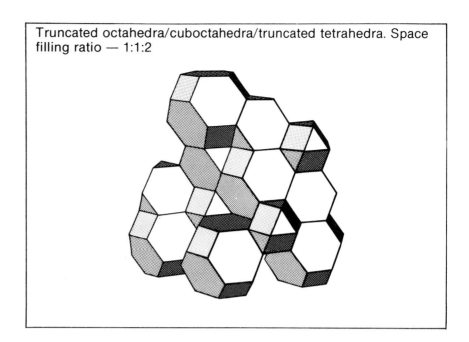
Truncated octahedra/cuboctahedra/truncated tetrahedra. Space filling ratio — 1:1:2

Uniform Space Filling with Four Kinds of Polyhedra

Regular/ Semiregular/ Prisms

Rhombicuboctahedra/truncated cubes/octagonal prisms/cubes. Space filling ratio — 1:1:3:3

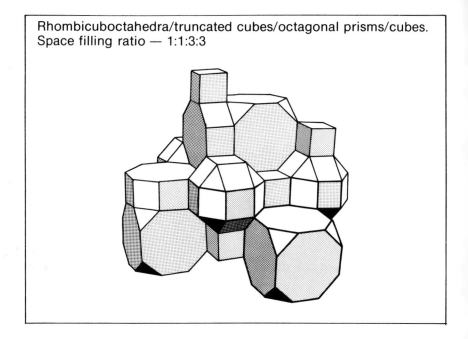

Uniform space filling systems, in which two or more kinds of semiregular prisms are combined, correspond to the semiregular tessellations. In order to visualize these systems refer to the illustrations of semiregular tessellations on pages 28–29.

Nonuniform Space Filling

In the case of the plane tessellations, there is an infinite number of nonuniform tilings with regular polygons. However, in the case of space filling with regular and semiregular polyhedra there are no nonuniform space filling systems. All of the possibilities of space filling with regular and semiregular polyhedra have been shown above. There can be an infinite number of nonuniform space filling systems, however, using other than regular and semiregular polyhedra.

Rhombic dodecahedra — the only Archimedean dual that will fill space alone.

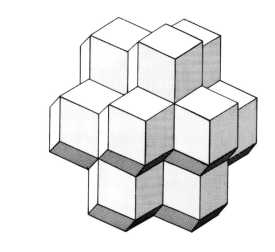

Trapezohedra

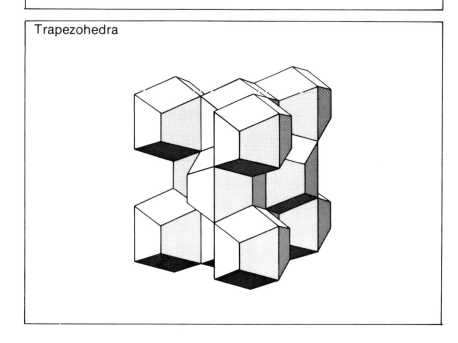

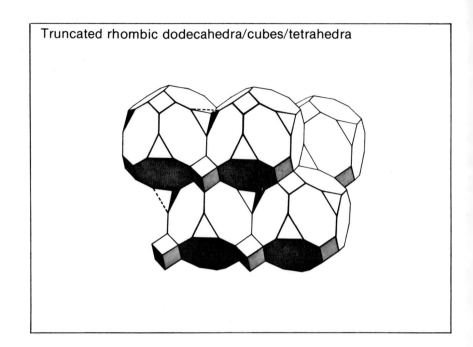
Truncated rhombic dodecahedra/cubes/tetrahedra

Dual Space Filling

A dual space filling network is formed by joining the center of each polyhedron in a space filling array with the centers of all of its neighboring polyhedra. A dual network will have as many different kinds of vertices as there are different polyhedra in the original space filling array.

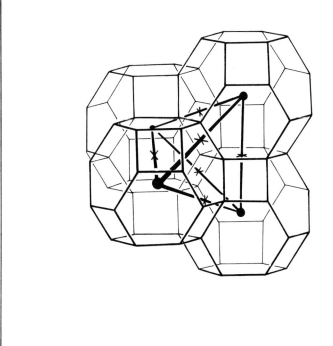

Self-Dual Space Filling

A space filling array of cubes is the only case in which the dual network consists of the same polyhedra as the original array. It is a self-dual.

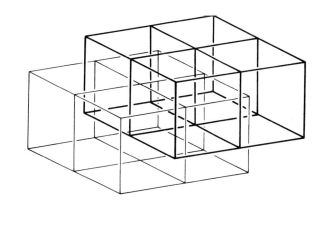

Complementary Prisms

The space filling systems of triangular prisms and hexagonal prisms are dual to each other.

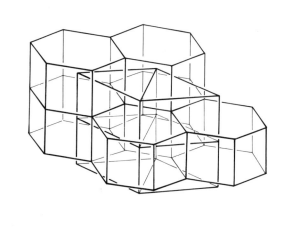

Polyhedra Formed by Dual Networks of Uniform Space Filling Systems

A tetragonal disphenoid is formed by the dual network of the space filling of truncated octahedra. This dual network is uniform.

Rhombic dodecahedra are formed by the dual network of the space filling of tetrahedra/octahedra.

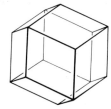

Rhombic hexahedra are formed by the dual network of the space filling of tetrahedra/truncated tetrahedra.

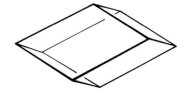

Square pyramids are formed by the dual network of the space filling of octahedra/truncated cubes.

Tetragonal octahedra are formed by the dual network of the space filling of octahedra/cuboctahedra.

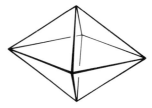

Quadrirectangular tetrahedra are formed by the dual network of the space filling of truncated cuboctahedra/octagonal prisms.

Right isosceles triangular prisms are formed by the dual network of the space filling of octagonal prisms/cubes.

Rhombic prisms are formed by the dual network of the space filling of triangular prisms/hexagonal prisms (2:1).

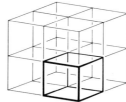 

Acute pentagonal prisms are formed by the dual network of the space filling of triangular prisms/hexagonal prisms (8:1).

Birectangular pentagonal prisms are formed by the dual network of the space filling of triangular prisms/cubes (2:1).

Obtuse pentagonal prisms are formed by the dual network of the space filling of triangular prisms/cubes (2:1).

Isosceles prisms are formed by the dual network of the space filling of triangular prisms/dodecagonal prisms.

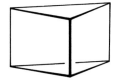

Trirectangular tetrahedra are formed by the dual network of the space filling of truncated cuboctahedra/truncated octahedra/cubes.

Right triangular pyramids are formed by the dual network of the space filling of truncated cuboctahedra/truncated cubes/truncated tetrahedra.

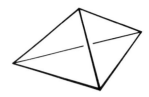

Trirectangular dipyramids are formed by the dual network of the space filling of rhombicuboctahedra/cuboctahedra/cubes.

Trigonal dipyramids are formed by the space filling of rhombicuboctahedra/cubes/tetrahedra.

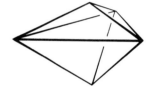

Rhombic pyramids are formed by the dual network of the space filling of truncated octahedra/cuboctahedra/truncated tetrahedra.

Birectangular quadrilateral prisms are formed by the dual network of the space filling of hexagonal prisms/triangular prisms/cubes.

Right triangular prisms are formed by the dual network of the space filling of hexagonal prisms/dodecagonal prisms/cubes.

Right square pyramids are formed by the dual network of the space filling of rhombicuboctahedra/truncated cubes/octagonal prisms.

6 Open Packings

Open Packings Derived from Space Filling Systems

Open packings are arrays of polyhedra in which all space is not filled. Some open packings are derived from space filling systems such as the truncated octahedra/truncated cuboctahedra/cubes. In this case, the open packing is formed by omitting the truncated cuboctahedra.

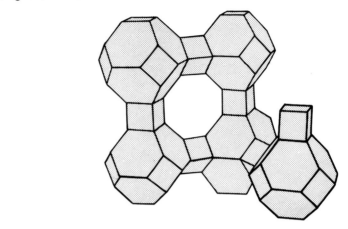

Open Packings Derived from Non-Space Filling Systems

Open packings can also be directly assembled as non-space filling arrangements of polyhedra. In this example of the open packing of hexagonal prisms and truncated octahedra, the prisms are attached to 4 out of 8 hexagonal faces on the truncated octahedra.

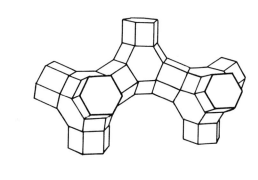

In this example of hexagonal prisms and truncated octahedra, the prisms are attached to all 8 hexagonal faces of the truncated octahedra.

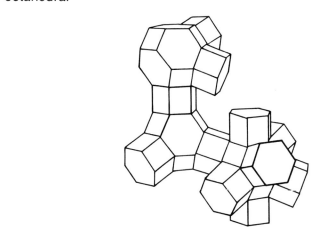

Truncated tetrahedra and hexagonal prisms. The hexagonal prisms are attached to all 4 hexagonal faces of the truncated tetrahedra.

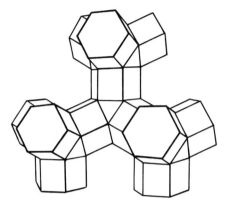

Truncated cuboctahedra and hexagonal prisms. The prisms are attached to all hexagonal faces.

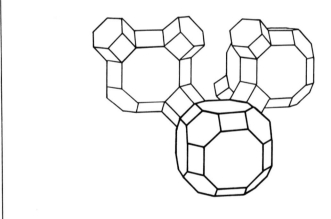

Open Packings of Tetrahedra, Octahedra, and Icosahedra

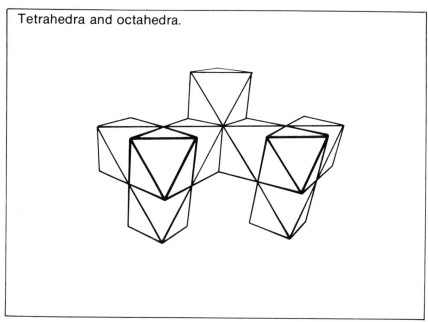
Tetrahedra and octahedra.

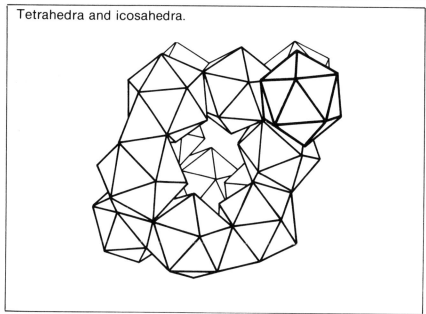
Tetrahedra and icosahedra.

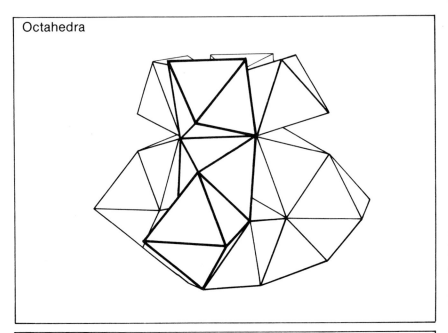
Octahedra

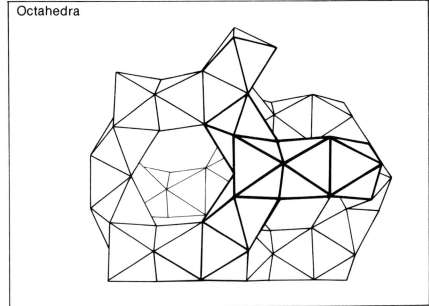
Octahedra

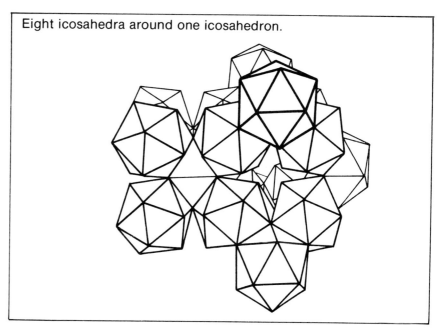
Eight icosahedra around one icosahedron.

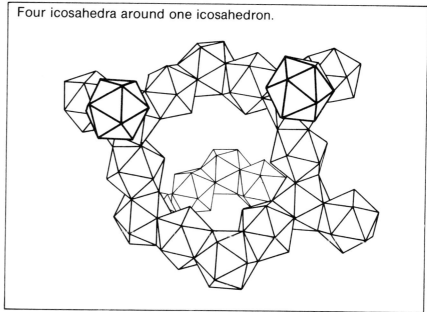
Four icosahedra around one icosahedron.

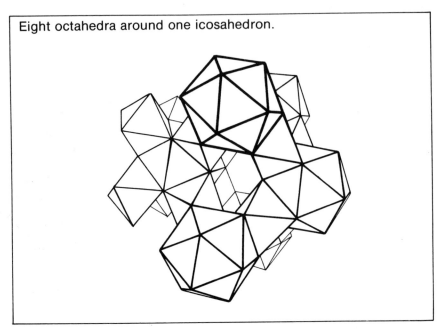

Eight octahedra around one icosahedron.

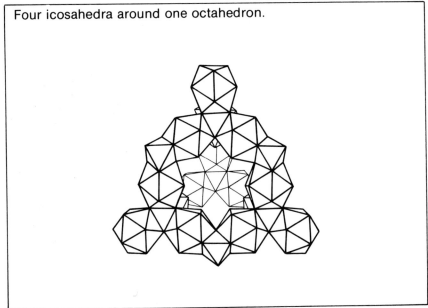

Four icosahedra around one octahedron.

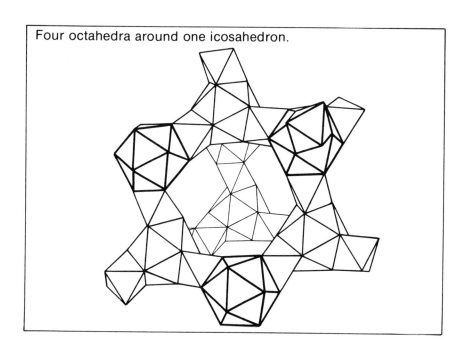
Four octahedra around one icosahedron.

7 Constructions

Basic Constructions

The characteristics of polyhedra cannot be adequately conveyed in two dimensions. The physical manipulation of actual three-dimensional polyhedra models is an important part of understanding them. An accurate polyhedron model begins with properly constructed faces or polygons. In order to draw polygons, you will need to understand some basic geometric constructions. A compass and a straight-edge ruler are the only items of equipment you will need for the basic constructions and for the construction of the polygons.

Bisecting a Line

Draw a line of desired length. Place the compass on one end of the line and swing an arc above and below the line. Place the compass on the other end of the line and repeat the procedure until the arcs intersect. Connect the points made by the arcs. The new line marks the point of bisection of the original line.

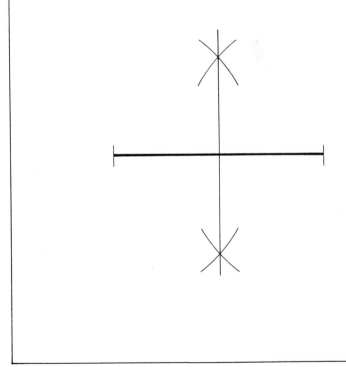

Bisecting an Angle

From the vertex of the angle, swing an arc AB across the angle. From point A swing an arc and repeat the procedure from point B until the arcs intersect at C. Connect point C with the vertex.

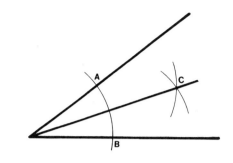

Constructing Perpendiculars

To a line at a given point in the line: From desired point A, swing an arc BC. From point B swing an arc. Repeat the procedure from point C until the arcs intersect at D. Connect points A and D.

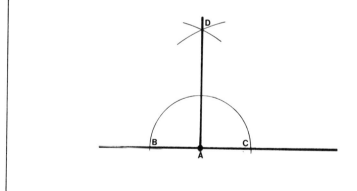

To a line through a point not in the line: From point A above the line, swing an arc BC below the line. From point B swing an arc below the line. Repeat the procedure from point C until the arcs intersect at D. Connect points A and D.

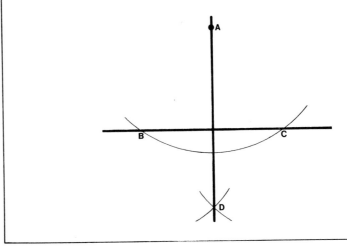

Constructing Polygons

Equilateral Triangle

Draw a straight line and mark off the desired edge length. Spread the compass the same edge length and swing intersecting arcs from each end of the line. Connect the points.

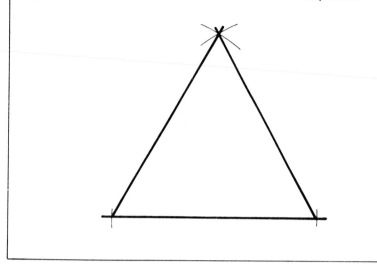

Right Triangle

Construct two lines perpendicular to each other. Measure desired edge length along each line and connect the points. If the two sides of the right triangle are equal, a right isosceles triangle is formed.

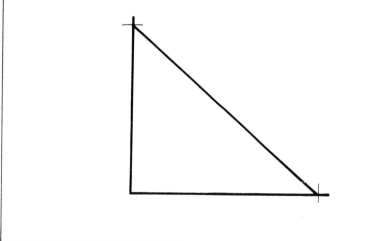

Isosceles Triangle

Draw a straight line and mark off desired length of base. Spread compass desired length of sides and swing intersecting arcs from both ends of the base. Connect the points.

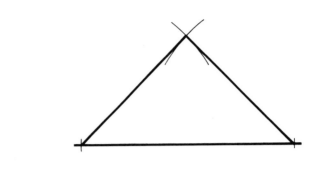

Square

Draw two lines perpendicular to each other and mark off desired edge length along both lines. Spread the compass the same length. Swing intersecting arcs from the ends of each side and connect the points.

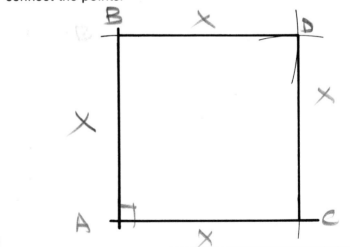

CONST
1. AB ⊥ AC
2. SET COMPASS TO X
3. SWING X FROM A THROUGH B & C
4. SAME COMPASS SETTING. SWING X FROM INTERSECTION OF ARC IN 3 AT B & C TO INTERSECT @ D
5. CAN ADD PROOF OF SQUARE BY USING COMPASS TO SEE THAT DIAGONAL AD = BC

Square Inscribed in a Circle

Draw two lines perpendicular to each other. With their point of intersection as the center, draw a circle with the compass. Connect the points where the lines intersect the circle.

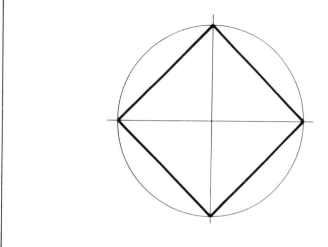

Regular Pentagon

Draw a circle with horizontal diameter CD, vertical diameter AB, and center 0. Bisect line OC and call it E. With E as the center and EA as the radius, draw an arc cutting OD at F. Line segment AF will equal one side of a regular pentagon. Swing successive arcs of the dimension AF along the circle. Connect the points.

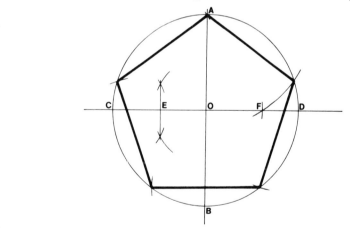

Regular Hexagon

Draw a circle of radius X. The radius equals one side of a regular hexagon. Draw successive arcs of the length X. Connect the points.

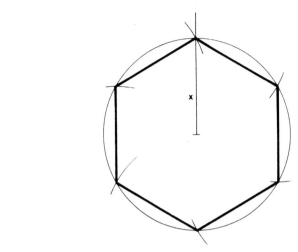

Regular Octagon

Construct a square in a circle. Bisect the central angles of the square. Connect the vertices of the square and the points of bisection where they intersect the circle.

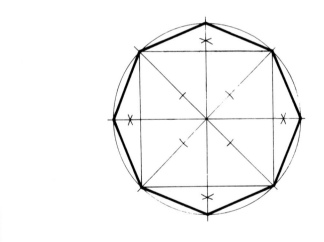

Regular Decagon

Construct a regular pentagon in a circle and draw and bisect the central angles. Connect the vertices of the pentagon and the points of bisection where they intersect the circle.

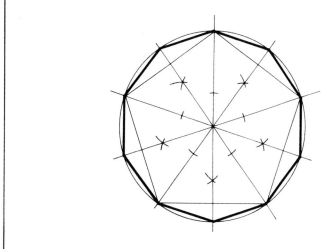

Regular Dodecagon

Construct a regular hexagon in a circle and draw and bisect the central angles. Connect the vertices of the hexagon and the points of bisection where they intersect the circle.

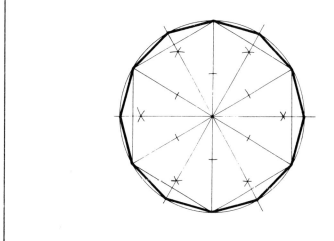

Modeling Polyhedra

Accuracy is important in the construction of polyhedra models. Also it is important that your hands are clean since the model materials can easily pick up smudges and fingerprints. The materials you will need are an X-acto knife with a Number 11 blade, a metal straight-edge, a cutting surface, and either cover stock, lightweight cardboard, tag board, or rigid plastic sheet. Further discussion of the type of model materials will follow in particular sections.

Making a Template

Construct the required polygons on lightweight cardboard or tag board. Make sure that for any given polyhedron the edge lengths of all the faces are the same. We find that a two-inch edge length makes a model of a convenient size. Construct the polygons as precisely as possible, since their accuracy will determine the precision of the finished models.

Carefully cut out the polygons with the X-acto knife and straight-edge. Make sure that the blade remains parallel to the edge of the straight-edge as you are cutting. Hold the straight-edge firmly so it will not slip since it serves as a guide for the cutting line. Using a slight pressure against the straight-edge, cut each edge with several sure strokes of moderate pressure until the edge is cut completely through. Always cut against a backing sheet, such as heavy chipboard, to avoid cutting table surfaces and to prolong the life of the knife blade.

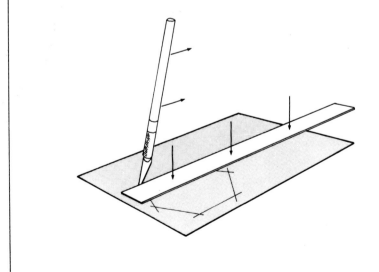

Paper Models

Using the templates, carefully trace the polygons on to the paper or cardboard you are using for your finished model. Carefully cut out the polygons as described above. We find that a white cover stock or tag board works best, resulting in a satisfying model in which the volume is easily perceived. Keep in mind that the cardboard should not be too thick as error will accumulate and the faces will not fit together precisely.

Scoring

Often it is easy to construct a polyhedron model by drawing a network and folding the faces to meet one another instead of cutting each face out individually. Lay out the network. Using a blunt instrument such as a ball point pen or other paper-scoring tool, and with the straight-edge as guide, score along the fold lines. Cut out the network along the outside edges and fold the faces until they meet.

Cube

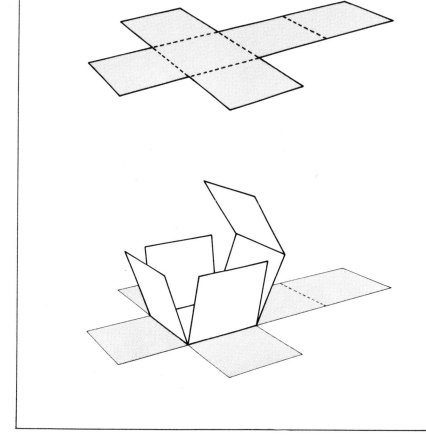

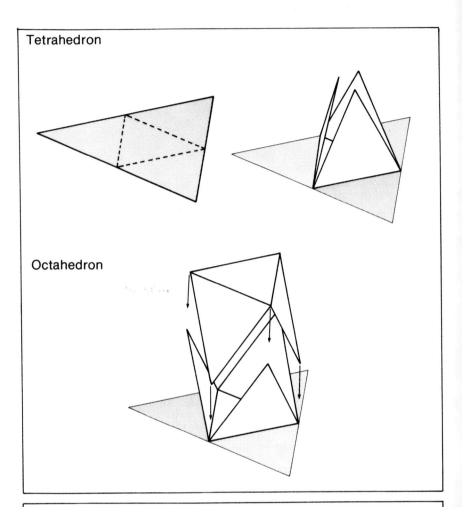

Plastic Models

Polyhedra models are especially attractive and durable when made from plastic. We recommend a rigid vinyl sheet plastic, .030" thick. The plastic comes in transparent or opaque. The transparent plastic results in models that are easy to perceive and understand in three dimensions. It is not practical to score and fold plastic models as you might with paper models. Therefore, each polygonal face must be cut out individually.

Using the knife and straight-edge, score the plastic by making several passes with the knife blade along the polygon edge. Keep scoring until you have a cut that is approximately halfway through the plastic. Grasp the plastic firmly with both hands, and using a swift, sure motion, bend back and snap the plastic apart at the cut. The plastic will snap apart giving a clean, straight cut. Cut out the remaining polygons.

For paper models, the best tape to use is Scotch Brand Magic Tape, No. 810, 1/2" wide. For plastic models, use Scotch Brand Polyester Tape, No. 853, 1/2" wide. Make sure your hands are clean as the tape will pick up any dirt.

Taping

Cut off a piece of tape slightly longer than the edge length of the polygon. Lay half the tape on one edge of a polygon. Bring another polygon up to the tape and join the two. Using a blunt tool, burnish the tape. (Special burnishing tools are available at art supply stores.) Cut off the extra lengths of tape. A hingeable joint will be formed that will bend to the appropriate dihedral angle as the polyhedron takes shape. Continue in a logical sequence to join all the polygons to complete the polyhedron. Placement of the last polygon will require attention and a bit of finesse.

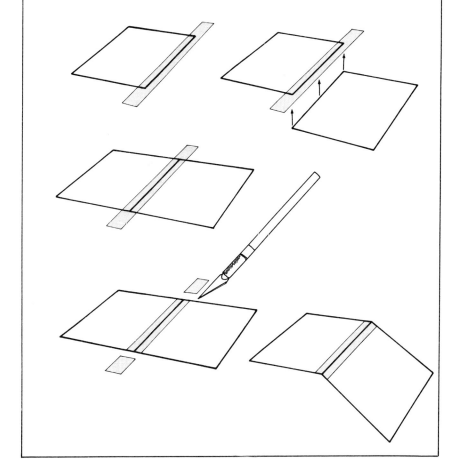

Bibliography

Avery, R.A. *Plane Geometry*. Boston: Allyn and Bacon, 1946.
This very old, high school geometry text was quite useful for geometric constructions and theorems. Though this particular book is probably unavailable, any good, current high school plane geometry text will be a useful reference.

Coxeter, H.S.M. *Regular Polytopes*. 2nd ed. New York: The Macmillan Company, 1963.
The authoritative mathematical treatment of spatial geometry—a classic in the field. It is a superb book, but difficult for non-mathematicians. Coxeter may be considered the foremost authority on the mathematics of polyhedra. Any work by him is of interest.

Cundy, H.M., and Rollett, A.P. *Mathematical Models*. 2nd ed. Oxford: Oxford University Press, 1961.
Probably the first popular book to contain a major section on polyhedra. Includes useful specifications for polyhedra and is easy to use for the nonspecialist.

Gardner, Martin. "On Tessellating the Plane with Convex Polygon Tiles." *Scientific American*, July 1975, pp. 112-114.
"More about Tiling the Plane: The Possibilities of Polyominoes, Polyiamonds and Polyhexes." *Scientific American*, August 1975, pp. 112-115.
"Extraordinary Nonperiodic Tiling that Enriches the Theory of Tiles." *Scientific American*, January 1977, pp. 110-121.
These three articles are an excellent and thorough discourse on the principles of tessellation.

Johnson, Norman W. "Convex Polyhedra with Regular Faces." *The Canadian Journal of Mathematics* 18:169-200 (1966).
This is the first exhaustive disclosure and proof of the existence of the 92 convex polyhedra with regular faces. Available in college math libraries.

Pearce, Peter. *Structure in Nature Is a Strategy for Design*. Cambridge, Mass.: The MIT Press, 1978.
An exploration of polyhedral structures in nature and their possible practical applications. Contains a thorough treatment of the geometry of space filling and an extensive bibliography.

Tuma, Jan J. *Technology Mathematics Handbook*. New York: McGraw-Hill Book Company, 1975.
A good reference for formulas and definitions. Any similar math handbook or dictionary can also serve the same purpose.

Wenninger, Magnus J. *Polyhedron Models*. Cambridge: Cambridge University Press, 1971.
An interesting presentation of individual polyhedra from the Platonic to the most elaborate stellated forms. Includes layouts for their construction and good discussions of theory.

Index

Angle, 2; bisecting, 119; dihedral, 46; interior, 5; measurement of, 2; types of, 3
Antiprism, 66; duals of, 84
Archimedes, 55

Bisecting: a line, 118; an angle, 119

Compass, 118
Concave: polygon, 6; polyhedron, 47
Congruent polygons, 6
Constructions, 117-129
Convex: polygon, 5; polyhedra with regular faces, 68; polyhedron, 47
Cube, 52; dual to octahedron, 75; in space filling, 90, 95-96, 98, 100
Cuboctahedron, 58; dual of, 80; in space filling, 93, 95, 97

Decagon, 8; construction of, 125
Decahedron, 49
Degrees, 2
Deltahedra, 67
Dihedral angle, 46; and space filling, 88

Dodecagon, 8; construction of, 125
Dodecahedron, 49; as dual, 76; as regular polyhedron, 53; stellation of, 70-71
Dual: space filling, 101; tessellations, 37; tilings of semiregular tessellations, 39-40
Duals: of polyhedra, 75-76; of semiregular polyhedra, 77-83
Dual polyhedra, 73-84; definition of, 74

Enantiomorph, 6
Enneagon, 8
Euler's Theorem: for polyhedra, 47; for tessellations, 44

Hexagon, 7; construction of, 124
Hexahedron, 48
Hexakaidecahedron, 49
Hexakis icosahedron, 83
Hexakis octahedron, 82

Icosahedron, 50; as dual, 76; as regular polyhedron, 54; in open packings, 111, 113, 114-115; stellation of, 72
Icosidodecahedron, 50, 59; dual of, 80
Icosioctahedron, 50
Interior: of polygon, 5

Kepler-Poinsot Solids, 70-71

Line: connecting two points, 2; bisecting, 118

Minutes, 2
Models: paper, 127; plastic, 128; scoring, 127; taping, 129

Naming: polygons, 7-8; polyhedra, 48-50
Network, 88; uniform, 89
n-gon, 7
n-hedron, 48
Nonagon, 8
Nonregular polygons, 10; in tessellations, 34-36; tiling with, 33

Octagon, 8; construction of, 124
Octagonal prisms: in space filling, 94, 98

Octahedron, 48; as dual to cube, 75; as regular polyhedron, 53; in open packings, 111-112, 115; in space filling, 92-93
Open packings, 107-115; from non-space filling systems, 109; from space filling systems, 108; of tetrahedra, icosahedra, octahedra, 111-115
Open patterns: concentric, 43; periodic, 42

Parallel, 2
Parallelogram, 12
Pentacontahedron, 50
Pentagon, 7; construction of, 123
Pentagonal hexecontahedron, 81
Pentagonal icositetrahedron, 81
Pentahedron, 48
Pentakaidecahedron, 49
Pentakis dodecahedron, 79
Perpendicular, 119-120
Plato, 52
Polygons, 1-20; combining, 22; definition of, 4; in dual tessellation, 37-38; nonregular, 10, regular, 9; star, 20; subdividing, 19; truncating, 20
Polyhedra, 45-71; combining, 86; dual, 73-84; formed by dual networks of uniform space filling systems, 102-106; modeling, 126-129; quasiregular, 58; regular, 52-54; semiregular, 55-65; stellated, 69
Prism, 66; duals of, 84; in space filling, 91, 94, 98

Quadrilateral, 7; types of, 12

Ratios: in space filling, 92-98
Rectangle, 12
Regular polygons, 9; in open patterns, 42, 43; in tessellations, 26-32
Regular polyhedra, 52-54; in space filling, 90, 92-93, 95-96, 98
Regular tessellations, 27
Rhombic dodecahedron, 80; in space filling, 99
Rhombic triacontahedron, 80
Rhombicuboctahedron, 63; dual of, 82; in space filling, 95-96, 98
Rhombicosidodecahedron, 65; dual of, 83
Rhombus, 12

Self-Dual: space filling, 101; tetrahedron, 75; tessellation, 38
Semiregular polyhedra, 55-65; duals of, 77-83; in space filling, 90, 92-98
Semiregular tessellations, 28-29
Septagon, 7
Septahedron, 48
Snub cuboctahedron, 60; dual of, 81
Snub icosidodecahedron, 61; dual of, 81
Space filling, 85-106; definition of, 87; dual, 101; nonuniform, 99; uniform, 90-98
Square, 13; construction of, 122-123
Stella octangula, 69
Straight-edge, 118
Symmetry: and polygons, 14; and polyhedra, 51; and tessellations, 41; combinations of, 17-18; mirror, 14; rotational, 15

Template, 126
Tessellations, 21-44; and symmetry, 41; dual, 37-38; regular, 27; self-dual, 38; semiregular, 28-29
Tetragon, 7
Tetrahedron, 48, 52; as self-dual, 75; in open packings, 111; in space filling, 92-96
Tetrakaidecahedron, 49
Tetrakis hexahedron, 78
Tiling: periodic, 25; uniform, 24: with nonregular polygons, 33; with regular polygons, 26
Trapezium, 12
Trapezohedra, 84; in space filling, 99
Trapezoid, 12
Trapezoidal hexecontahedron, 83
Trapezoidal icositetrahedron, 82
Triakis octahedron, 78
Triakis icosahedron, 79
Triakis tetrahedron, 77
Triangle, 7; types of, 11; construction of, 122
Truncated cube, 56; dual of, 78; in space filling, 93, 96, 98
Truncated cuboctahedron, 62; dual of, 82; in space filling, 94-96
Truncated dodecahedron, 57; dual of, 79
Truncated icosahedron, 57; dual of 79
Truncated icosidodecahedron, 64; dual of, 83
Truncated octahedron, 56; dual of, 78, in space filling, 90, 95, 97
Truncated rhombic dodecahedra: in space filling, 100
Truncated tetrahedron, 55; dual of, 77; in space filling, 92, 96, 97

Uniform: network, 89; polyhedron, 47; space filling, 90-98

Vertices: typical with regular polygons, 26
Vertex: of polygon, 4; of polyhedron, 46; of tessellation, 24